THE 50 BEST SIGHTS IN ASTRONOMY AND HOW TO SEE THEM

Observing Eclipses, Bright Comets, Meteor Showers, and Other Celestial Wonders

Fred Schaaf

BICENTENNIAL 1807 WILEY 2007 BICENTENNIAL

John Wiley & Sons, Inc.

Dedicated with love and gratitude to my sisters,
Pat and Joanne, and my brother, Bob.

This book is printed on acid-free paper. ♾

Published by John Wiley & Sons, Inc., Hoboken, New Jersey
Published simultaneously in Canada

Wiley Bicentennial Logo: Richard J. Pacifico

Credits appear on page 273.

Design and composition by Navta Associates, Inc.

For general information about our other products and services, please contact our Customer Care Department within the United States at (800) 762-2974, outside the United States at (317) 572-3993 or fax (317) 572-4002.

Wiley also publishes its books in a variety of electronic formats. Some content that appears in print may not be available in electronic books. For more information about Wiley products, visit our web site at www.wiley.com.

Library of Congress Cataloging-in-Publication Data:

Schaaf, Fred.
The 50 best sights in astronomy and how to see them : observing eclipses, bright comets, meteor showers, and other celestial wonders / Fred Schaaf.
p. cm.
Includes bibliographical references and index.
ISBN 978-0-471-69657-5 (pbk. : alk. paper)
1. Astronomy—Observer's manual. 2. Astronomy—Amateurs' manual. I. Title.
II. Title: Fifty best sights in astronomy and how to see them.
QB63.S389 2007
520—dc22

2006036221

Printed in the United States of America

10 9 8 7 6 5 4 3 2 1

CONTENTS

ACKNOWLEDGMENTS

In the late 1980s and early 1990s, Wiley published four of my books. Now this book marks a return for me to Wiley. Kate Bradford is the editor whom I worked with on several of the past books and the person who acquired this book and my next, *The Brightest Stars* (due out from Wiley next year). I wish to thank Kate so much for playing this role and for being continually supportive and helpful over the years.

In the first phase of working on *The 50 Best Sights in Astronomy*, my editor was the congenial Teryn Johnson. In the next stage, my editor was Christel Winkler, who was patient and understanding under trying circumstances. I wish to give my deepest thanks to her for her tremendous and conscientious efforts to keep this book on schedule. Another person I want to thank for her hard and vital work on this book is editorial assistant Juliet Grames.

The photographers and artists who have contributed to this book are many, and their additions have been crucial.

The largest number of diagrams and maps were supplied by two old friends of mine: Guy Ottewell and Doug Myers. Their work is always unique and brilliant.

Vital maps were also provided by Robert C. Victor and D. David Batch, who produce the Abrams Planetarium Sky Calendar. The Sky Calendar is a wonderful resource for all knowledge levels of skywatchers as well as for teachers. It is available from the Abrams Planetarium, Michigan State University, East Lansing, Michigan 48824. You can also check out the associated Skywatcher's Diary at www.pa.msu.edu/abrams/diary.html.

The largest number of photographs by far were provided by Johnny Horne. I've enjoyed the originality of Johnny's amazing work for many years and relied on him for some comet images a decade ago. But it was this current book—and the 2004 transit of Venus—that got me back in touch with him.

An old friend who provided both photos and sketches was Ray Maher. Other good friends who were photo contributors include Steve Albers, Nelson J. Biggs Sr., and Chuck Fuller. Akira Fujii, Richard Yandrick, and Shahriar Davoodian supplied several of this book's most stunning images, for which I thank them.

INTRODUCTION

The idea of this book is really quite simple. I've chosen the fifty best sights based on my own lifetime of passionately seeking astronomical wonders. I try to present them here through vivid detail and concise observing instructions.

Nothing could be more clear and simple, nor, if you wish to seek the very heart of astronomy, any more desirable. This book offers to those seeking the thrills of the heavens the most instant and immediately powerful gratification of that desire possible from pages. It is, after all, astronomy honed down to the most exciting experiences. Armchair astronomers will find all they need to enjoy the heavens. Most people, however, will be motivated to action. They will, I hope, be inspired, more electrifyingly than would otherwise be possible, to get out and experience these wonders for themselves.

A remarkable thing about *The 50 Best Sights in Astronomy* is that no one seems to have ever had this idea for a book before. Why did I stumble on the idea when other people didn't? I'll offer my explanation in a moment.

First, I'd like to mention something else about this book that I think is original and that you may find interesting. I'm talking about the criterion I've used for determining the order in which the sights should be discussed. My criterion is the width of the field of view necessary to enjoy the sight best.

The order is from the widest to the narrowest view. The first sight is that of the entire starry sky on a clear night, preferably with no interfering moonlight or city light pollution. This is a scene at least 180° wide—the angular span of the entire dome of the sky. Of course, it's not really possible to be so precise as to determine that a slightly larger field of view is needed for enjoying a bright star in its stellar setting as compared to a bright planet in its own setting. So the book is divided into sections by width of field. The first section includes sights that require a field ranging from that of the entire sky down to 100° wide—a broad naked-eye scan. Each succeeding section zooms farther in until we start looking at the magnified but also much narrower views available through binoculars. And then the final section contains sights that are the most magnified and narrow of all the ones viewed through telescopes.

This ordering system is interesting not just because of its novelty. It makes sense because roughly the first half of the book includes sights that occur over a wide enough area of the sky to be best seen with unaided eyes (we're talking sights such as meteor showers, the Northern Lights, the Milky Way, and impressive constellations). Only as we get farther along in the book do binoculars and finally telescopes become necessary. Beginners will find that much of the book can be used even if they don't own any optical instruments (other than their own eyes). But most people who read a little about astronomy and see some of its sights soon find themselves wanting a telescope. So it's good for the latter part of the book to provide material for people who either have a telescope or suspect they may get one soon.

Besides providing a glossary, the book opens with a quick guide to the basics of observational astronomy that novice astronomers may wish to study before moving on to the fifty chapters about the sights. Throughout the book, you will find terms in boldface. These terms can be found in the glossary at the back of the book.

Furthermore, despite its emphasis on presenting the purely visible and experiential aspects of astronomy, this book's chapters also contain explanations of the sights. I have simply endeavored to keep these explanations concise and weave them seamlessly into the descriptions. The explanations can add much to a person's appreciation of sights that are already visually spectacular. For instance, the naked-eye view of Venus at its brightest is one of astronomy's best sights; it is tremendously impressive even without explanation or commentary. But when you learn *why* Venus is far brighter than any other planet and why sometimes it's even more soul-searingly brilliant than at other times, this knowledge can't help but increase your sense of wonder and your thrill in seeing Venus.

Of course, in *The 50 Best Sights in Astronomy*, the more intense and potent the verbal descriptions of the sights and what it is like to see them, the more powerful the book will be. And I believe that this is precisely where I am at my strongest as a writer and as an observer. I've seen these sights myself passionately through more than four decades of observing.

Now let me return to the question I posed earlier: Why hasn't anyone before thought to focus a book on just the most spectacular visible sights of astronomy?

I think the primary reason is the wide *variety* of objects and phenomena that need to be covered in such a book. Observing features in the Moon's landscape is very different from looking for wispy, faintly glowing galaxies, identifying constellations, sitting in a lawn chair counting meteors for hours, or keeping track of the brightness changes of variable stars over the course of weeks, months, and years.

An important device of mine in conceiving this book and in communicating it to you is the use of the inclusive term *sights.* Planets are objects or bodies; the Northern Lights a phenomenon; the Big Dipper a pattern; the Great Red Spot of Jupiter a cloud feature; a meteor shower a display—but all are sights. And the use of this term allows us to include certain great astronomical wonders that are not often focused on at all as discrete things: a clear, moonless, star-filled country sky; the winter grouping of constellations surrounding Orion; a single bright star seen above an impressive landscape (a person often sees and greatly admires such a star in the context of sky and landscape, but because we have no simple name for the entirety of what is being seen, it doesn't get written about).

Let me stress that I feel that *The 50 Best Sights in Astronomy* will appeal to the widest possible audience—not just experienced amateur astronomers but truly anyone who has ever looked up, or even considered looking up, at night. I think this book offers a completely original angle and format for presenting what's exciting in observational astronomy. And—as this book shows—what's exciting about observational astronomy can change your life.

Basic Information for Astronomical Observers

The following information is not necessary to know for many of the sights in this book. For some, however, it is very important.

Observing Conditions, Techniques, and Telescopes

Skies and Eyes

The most basic requirements for observational astronomy are obviously a clear sky and a set of eyes. But much about what constitutes a clear sky is not obvious, and there is much for a beginner to learn about exactly *how* to use those eyes.

To most people, a clear sky simply means one without clouds. But astronomers speak of **transparency**—the amount of light the atmosphere can pass. Two nights could be cloudless. Yet one night could be plagued with a thick haze composed of much water vapor (high humidity), human-made air pollution, or natural particles such as dust and pollen. Through this haze, only a few stars are visible. But the other night could offer an atmosphere nearly devoid of these ingredients and, through this transparent atmosphere, thousands of vividly shining stars visible in a velvet black sky.

The two nights just described are assumed to occur at a country site far from city lights. But even if you have a very transparent atmosphere, your sky may be glaringly bright and show few stars if it is severely afflicted by **light pollution**. Light pollution is society's excessive and misdirected outdoor lighting. It is by far the greatest threat to most of amateur astronomy and to everyone's enjoyment of the beauty of the night sky. My discussion of it and what you can do about it appears in Sight 1.

Before we turn from skies to eyes, we should note that it is not just transparency that astronomers hope for in a night sky. They also want good **seeing**. *Seeing* is the measure of how steady the atmosphere is. The more turbulent the atmosphere, the less crisp your images in a telescope will be. Sometimes even your unaided eye can tell that seeing is bad by noticing that there is unusually strong twinkling of the stars, even high in the sky.

Now, what about your eyes? To use them best you need to know about dark adaptation and averted vision.

Dark adaptation is a chemical process that occurs in the eyes when they are in a low-light environment. It takes about 15 to 20 minutes for the process to make your eyes much more sensitive to faint light (there is continued slight improvement the longer you are in the dark). The key thing to remember is that the reintroduction of a fairly bright light—such as someone's flashlight pointing up into your eyes—will spoil your dark adaptation and force you to begin the process all over again. Interestingly, red light does little to spoil dark adaptation. For that reason, many amateur astronomers put red filters of some kind on their flashlights (such flashlights are commercially available from some companies that sell telescopes and other astronomy equipment, but you can make your own with a piece of red cellophane).

Averted vision is looking just to the side of a faint star or other faint object so as to bring its light on the most sensitive parts of your retina. Try this technique—it can enable you to see sights several times dimmer than you can with direct vision!

Basics of Binoculars and Telescopes

Many great astronomical sights can be seen—some even best seen—with the naked eye. But most amateur astronomers eventually want to get a telescope.

A preliminary step may be to buy a pair of *binoculars.* These are really paired and coordinated small telescopes that have the virtues of being very portable and offering a wider field of view than telescopes. A statistic like 7×50 when applied to binoculars means it makes things look 7 times bigger (7×) and has primary lenses that measure 50 mm across.

Binoculars typically offer much less magnification than telescopes. But in astronomy, light-gathering power is generally far more important than magnification. The light-gathering power and the resolving ability (ability to show fine detail) of a telescope depends on the size of its main lens or mirror—all else being equal, the wider that piece of glass, the better.

The most basic parts of any telescope are its *mount* (which should be rock-steady), its *tube* (though some telescopes use an open series of trusses or other open design), and its optics—which include a *primary mirror* or *lens.* An *ocular* or *eyepiece* is a small assemblage of lenses that can be used with different telescopes to magnify the image sent from the telescope's optics.

There are three most basic types of telescopes. A *reflector* uses mirrors and usually puts the primary mirror at the bottom of the telescope and a secondary mirror and eyepiece holder at the top (skyward end). A *refractor* uses lenses and puts the primary lens at the top (skyward end) and the eyepiece holder at the bottom. A *catadioptric* uses both mirrors and lenses.

Fundamental Measurements in Astronomy

Positions in the Sky

An astronomer needs to find his or her way around the night sky. The first step is learning about two coordinate systems.

The **altazimuth system** uses altitude for vertical measure and azimuth for horizontal measure. Altitude is measured in degrees from 0° (the horizon) to 90° (the zenith—the overhead point in the sky). Azimuth is measured in degrees from 0° (due north) around to 90° (due east) and so on around the sky to 360°—really the 0° of north again.

The **equatorial system** is based on the concept of a **celestial sphere** that is imagined to be positioned around Earth with a north celestial pole over the north geographic pole, a celestial equator over the geographic equator, and a south celestial pole over the south geographic pole. **Declination** on the celestial sphere is the counterpart of latitude on Earth but is measured in positive degrees (0° to +90°) north of the celestial equator and negative degrees (0° to –90°) south of the celestial equator. Each degree of declination can be further divided into 60 minutes (written as 60') of declination and each minute into 60 seconds (written as 60") of declination. **Right ascension** (RA) on the celestial sphere is the counterpart of longitude on Earth, but instead of using a 0° line that forms the prime meridian running through Greenwich, England, it uses a 0 h (zero hour) line that runs through the vernal equinox point in the sky (the point the Sun is at in the sky at the start of spring in Earth's northern hemisphere). Also, instead of RA being measured in degrees, 180° West from the prime meridian and 180° East from the prime meridian as longitude is, RA is measured in 24 *hours* of right ascension east around the celestial sphere. Each hour of right ascension is further divided into 60 *minutes* (written as 60 m) of right ascension and each minute into 60 *seconds* (written as 60 s) of right ascension.

Angular Distances

Observers also have a need to determine how many degrees it is from, say, a star to a galaxy in the sky or how many degrees long a comet's tail appears. Here, we can divide each degree into 60 arc-minutes (written as 60') and each arc-minute into 60 arc-seconds (written as 60"). And we can do a rough estimate of angular measure in the sky by holding out our fist or finger at arm's length. The width of a person's fist at arm's length is about 10°, the width of a little finger at arm's length about 1½°. In comparison, the Big Dipper is more than 25° long, the Moon is about ½° (30") wide, and Venus, when it appears largest, is about 1' (60") wide.

Distance in Space

The most famous unit of measure for interstellar and intergalactic distances is the **light-year**. It is the distance that light (or any form of electromagnetic radiation), the fastest thing in the universe (at about 186,000 miles per second or 300,000 kilometers per second), can travel in one year. A light-year is about 6 trillion (that is, 6 million million) miles.

Brightness

In astronomy, brightness is measured by **magnitude**. Originally, all naked-eye stars were categorized in six classes of brightness, from 1st magnitude (brightest) to 6th magnitude (faintest). In modern times, the scale has been extended to zero and to negative magnitudes for very bright objects (remember, the lower the magnitude, the brighter the object). It has also been extended to numbers higher than 6 for objects so faint they require optical aid to see. Decimals are used between two magnitudes: a star midway in brightness between 1.0 and 2.0 is 1.5. A difference of one magnitude means one object is about 2.512 times brighter than the other. This is because it was considered useful to set a difference of five magnitudes as equal to 100 times—2.512 (actually 2.512 . . .) multiplied by itself 5 times is 100.

A Note on Stellar Nomenclature

There are various names and designations that are given to stars. Many of the naked-eye stars receive a scientific name using a letter of the Greek alphabet and the genitive form of the constellation in which they are located—for instance, Alpha Leonis, the alpha star of the constellation Leo. The *Bayer letters* (so called because they were mostly applied to the stars by the astronomer Johannes Bayer) often work in order of decreasing brightness (the first letter of the Greek alphabet—alpha—is given to the brightest star in the constellation). But variations from this scheme are common (for example, Alpha Orionis is the second brightest star in Orion, not the brightest).

The Scheme of the Universe

The solar system consists of the Sun—the star closest to us—and objects that are all orbiting around the Sun (though in the case of moons, indirectly). These objects are planets and their moons, asteroids, meteoroids, comets, and Kuiper Belt Objects. Earth orbits or revolves around the Sun, as do the other planets. Earth is in turn orbited by the Moon, and most of the other planets by moons of their own. **Asteroids** are rocky worlds that mostly circle the Sun between the orbits of Mars and Jupiter and are much smaller than planets. **Meteoroids** are rocky bodies smaller than asteroids, roughly less than a few hundred feet in diameter. **Comets** are predominantly icy objects that typically follow elongated orbits that bring them much closer to the Sun at one end of their orbit—at which point they are heated enough to release clouds of gas and dust to form a head and often a tail. *Kuiper Belt Objects* are mixtures of rock and ice orbiting mostly beyond the orbit of Neptune.

Countless **stars** are found far beyond the orbits of our solar system's most distant planets and comets. Stars come in a number of varieties—blue giants, red giants, white dwarfs, red dwarfs, and various kinds of "main sequence" stars are examples. Many stars are **double stars**—on closer examination, one point of light turns out to be two, two suns that may be orbiting each other or at least going through space together. Many stars are **variable stars**—that is, they vary in brightness, regularly, semiregularly, or irregularly. The most spectacular kinds of variable stars are exploding stars of two sorts: novae (in which a certain small percentage of the star blows away) and supernovae (in which, by a completely different process, 80 percent or more of the star blows away). A star with a mass similar to the Sun goes from being a main sequence star to a red giant to a white dwarf. A star with much less mass than the Sun becomes (or already is) a red dwarf for life. A star with many times more mass than the Sun goes supernova and collapses into a star much smaller than a white dwarf, a **neutron star** (which may also be a **pulsar**). A star with very many times the mass of the Sun collapses into a star so small it "has no size at all"—a **black hole**. A supernova may result in a cloud of gas that lingers in sight for hundreds or thousands of years—a **supernova remnant** (SNR).

There are two types of **star clusters**. An **open cluster** (also called galactic cluster) is fairly loose in its arrangement of stars and consists of typically a few dozen to many hundreds of stars. A **globular cluster** is a tight concentration of stars that consists of typically tens of thousands or hundreds of thousands of stars.

A **nebula** is a cloud of gas and dust in interstellar space. A **diffuse nebula** is a place where new stars are being born. It may be either an *emission nebula* (shining on its own after being stimulated by extremely hot stars) or a *reflection nebula* (shining by virtue of reflecting the light of nearby stars). Another kind of nebula is a **planetary nebula** (see Sight 49 for an explanation of the name), a cloud of gas and dust released from a dying star.

The stars we see in the sky with our naked eyes are all part of a much vaster congregation of stars called a **galaxy**. Our galaxy is called the Milky Way (a name also given to the band of light its star-crowded equatorial disk produces in our sky). The Milky Way contains a few hundred billion stars.

Beyond our Milky Way Galaxy exist innumerable other galaxies: *spiral galaxies, elliptical galaxies,* and *irregular galaxies.* There are also gatherings of galaxies—such as the Local Group of galaxies—to which we belong.

Among very distant galaxies are powerful beacons of light, radio waves, and other radiation called **quasars**. If cosmologists are correct, all galaxies beyond our gravity-bound Local Group are receding from us—the farther they are, the faster the recession. This recession is caused by the continuing expansion of space itself in the universe.

Field of View

180°

(THE WHOLE SKY)

TO

100°

(NAKED-EYE SCAN)

THE STARRY SKY

The Sun, the Moon, and the stars. These are the three greatest, most elemental sights in astronomy.

Unfortunately, we have seen the Sun and the Moon on so many days and nights of our life that we have foolishly learned to take them for granted. We wake up only when something like a solar eclipse (see Sights 2, 19, and 38), a lunar eclipse (Sights 18 and 38), or a view of lunar mountains and craters in a telescope (Sights 35 and 36) remind us of the awesomeness of these orbs.

But what about the stars? A view of a single bright star in all its splendor is one of the best sights in astronomy (especially if the star is the brightest of all night's stars—see Sight 14). But I would argue that the greatest astronomical sight is not a mighty comet that stretches a quarter of the way across the sky for a few weeks (Sight 13), not a fireball that turns night into day for a few seconds (Sight 4), not even a total eclipse of the Sun that turns day into night for a few minutes (Sight 2). It is the very heart of night and the wellspring of all night's wonder: the starry sky.

Astronomy's Greatest Sight

Of course, what I mean is not just a glimpse of the starry sky, seen between trees, or the glaring light-polluted sky that stretches over large cities today. I mean the whole starry sky, seen on a clear, moonless night many miles from city lights, seeming to curve over us as a celestial dome. In other words, I mean the night sky in its natural state, seen as it should be.

This view of the entire starry sky is the grandest sight in the universe. How could it not be? It is, after all, our sight *of* the universe—the largest physical view we can ever have. Of course, only about half the universe is above our horizon at one time if we observe it from the surface of the Earth. But this hemi-universe contains in it examples of every astronomical wonder that humankind knows—along with our landscape and ourselves.

The Dark Sacred Night

Under the best conditions, a few thousand stars can be seen directly in the sky with the naked eye, and thousands more indirectly (that is, using averted

LEGEND
Star Magnitudes

- Zero or brighter
- 1st
- 2nd
- 3rd
- 4th
- 5th
- * Deep Sky Objects

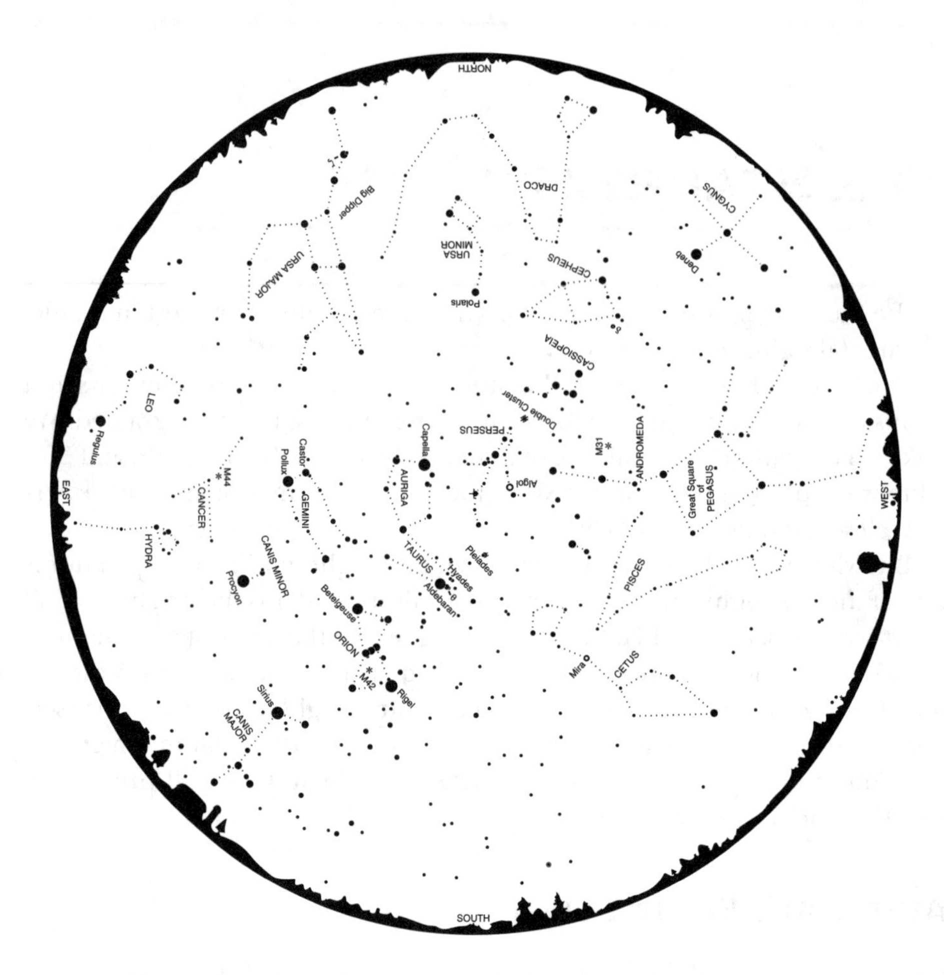

This chart is drawn for latitude 40° N but is useful throughout the continental United States. It represents the sky in mid-January at 8:30 P.M. local standard time and is applicable one hour either side of this time. For each earlier month, go out two hours later, and vice versa.

vision to see much fainter sights, as discussed in the introduction). Every one of these points of light seems to exist only to send illumination to all corners of the universe. Every one seems to tremble as if it has a life of its own. But the realities behind these charming fancies about the stars are so amazing that they seem even more fantastical—and yet are scientifically true. Which realities do I mean? The reality that every one of these points is a sun of its own—just a sun that appears tremendously dimmer than ours because it is millions of times farther away than our Sun. The reality is that many of these points of light possess planets, and that some of these planets probably harbor life, possibly even intelligent life.

Whatever your views on religion are, you may recall and appreciate a line from the Louis Armstrong song "What a Wonderful World." The line salutes equally "the bright blessed day" *and* "the dark sacred night." Readers of the Bible will recall that the book of Genesis says that "on the first day of Creation God separated darkness from light, calling one Day and the other Night."

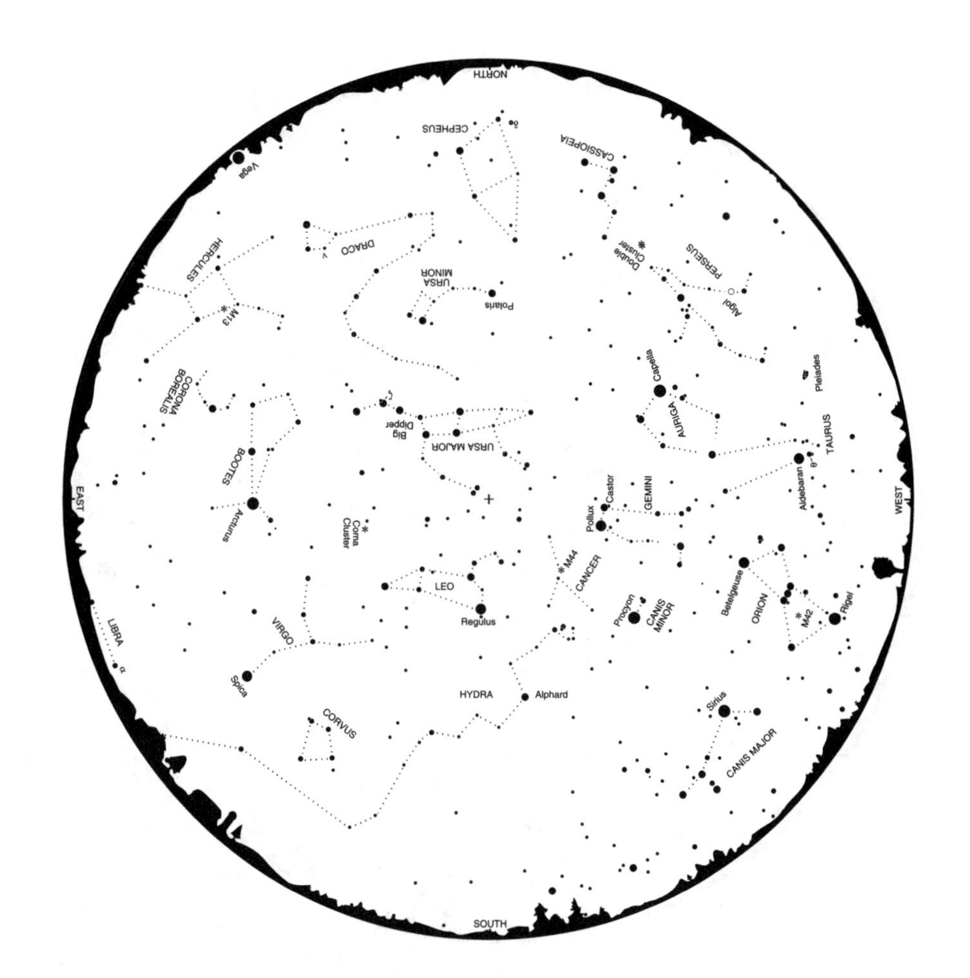

This chart is drawn for latitude 40° N but is useful throughout the continental United States. It represents the sky in mid-April at 9:30 P.M. local daylight saving time and is applicable one hour either side of this time. For each earlier month, go out two hours later, and vice versa.

Save Our Greatest Wonder

Sadly, however, the human race is now throwing up a wasteful glare about itself that is preventing most of us from seeing this "face" of splendor and ultimate meaning, the starry sky. Our greatest wonder of astronomy, our most inclusive and comprehensive view of the universe, is one that is now, in our time, threatened. The enemy is **light pollution**—that component of human-made outdoor lighting that is excessive and misdirected. Light pollution does no one any good, and whether you are a skywatcher or not, you should care about some of the adverse practical effects of light pollution. Light pollution wastes billions of dollars a year worth of electricity just in the United States, leads to increased fossil fuel burning that has numerous negative effects on the environment, and puts dangerous glare in the eyes of both motorists who are trying to avoid accidents and homeowners who are trying to detect intruders. Light pollution has a fatal impact on many kinds of animals and may possibly have some very deleterious effects even on human health.

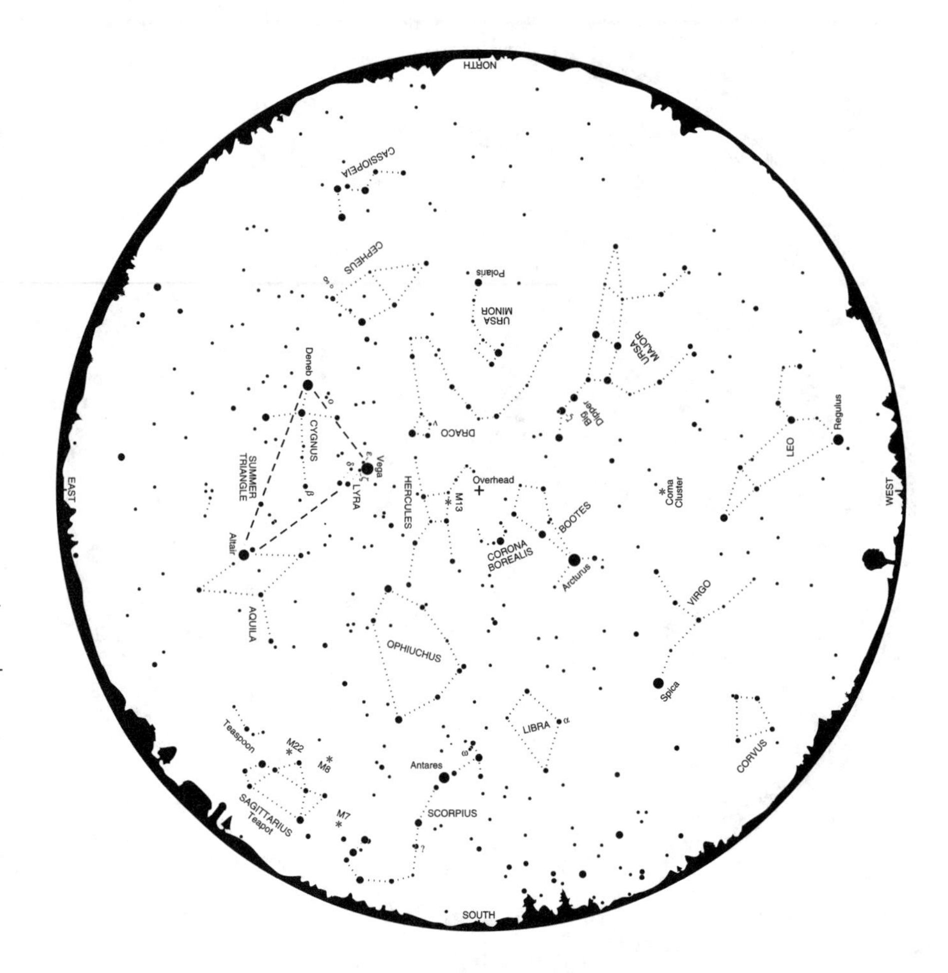

This chart is drawn for latitude 40° N but is useful throughout the continental United States. It represents the sky in mid-July at 9:30 P.M. local daylight saving time and is applicable one hour either side of this time. For each earlier month, go out two hours later, and vice versa.

How bad has light pollution gotten in the past few decades? A 2001 study used satellite imagery of city lighting to produce the first light pollution atlas of the world (www.lightpollution.it/dmsp/index.html). This atlas shows that about two-thirds of the world's population now live under light-polluted skies and that 99 percent of the people in the United States and the European Union do. About 20 percent of the world's population and more than two-thirds of the U.S. population can no longer see the Milky Way from their homes (see Sight 7 for a taste of what so many of us are missing). About 10 percent of the human race live where light pollution is so severe that it never even allows their eyes to dark-adapt. This figure for U.S. residents is an astonishing 40 percent.

Fortunately, about three-fourths of all light pollution can be eliminated with practices and technology that already exist and are affordable—indeed that ultimately will save billions of dollars. The first step is to go to www.darksky.org, the marvelous Web site of the International Dark-Sky Association (IDA), an organization that now has more than 10,000 members.

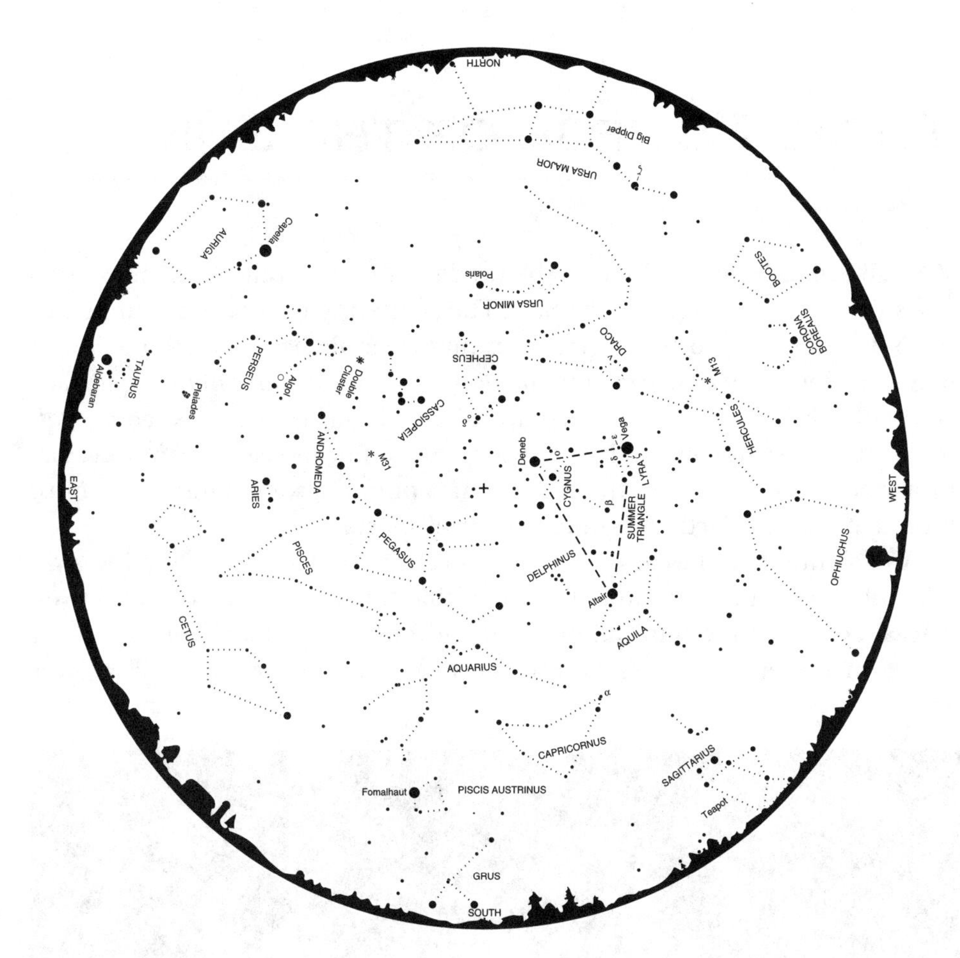

LEGEND
Star Magnitudes

- ● Zero or brighter
- ● 1st
- • 2nd
- • 3rd
- · 4th
- · 5th
- * Deep Sky Objects

This chart is drawn for latitude 40° N but is useful throughout the continental United States. It shows the sky in mid-October at 9:30 P.M. local daylight saving time and is applicable one hour either side of this time. For each earlier month, go out two hours later, and vice versa.

More Stars Than Sky

Thanks to the IDA and dark-sky activists, there really has been some progress in many locales in the battle against light pollution. Of course, much more work is needed. The light pollution from very large cities can often be detected from more than 100 miles away. Even a small city of 20,000 may appear as a sizable glow on your horizon 15 miles away. But at least some of the factors that determine our enjoyment of the night sky are relative—relative to what we are used to. A case in point was the experience of a friend of mine. I rate the darkness of a clear night sky on a basic 1 to 7 scale, in which 1 is the sky from within a typically light-polluted big city and 7 is the sky at a pristinely dark site. My friend spent most of his adult life in a level-2 or -3 sky. But one week he went camping at a more remote location in the mountains. I believe he probably experienced no darker than a level-5 sky. What was his reaction? It was awe. “At midnight,” he said, “there were more stars than sky.”

Sight 2

Total Eclipse of the Sun

Of all events in nature that can be predicted far in advance, the most awesome is a total eclipse of the Sun. The shattering beauty and strangeness of a total solar eclipse are so great that many people have something almost like a religious experience during the period when the Sun's blinding surface is completely hidden by the dark shape of the Moon. That time is called *totality*. It never lasts for more than a few minutes. But in years when it occurs, thousands of people often travel thousands of miles, sometimes to the most remote places on Earth, to witness the brief spectacle.

A small number of total solar eclipses occur each decade, but totality is visible only from a narrow band of territory that the Moon's shadow races over. A total eclipse of the Sun will occur at a given spot on Earth an average of about once every 375 years. So travel is necessary for most of us if we ever

The diamond ring at a solar eclipse.

want to see "the coming of the sudden shadow" and have the experience of Moon-made midday darkness.

I hope now to convince the reader who knows little about total eclipses of the Sun to commit to making the trip to totality at some time in the years ahead. What follows is a description of what a particular total solar eclipse might be like, based partly on those that I have personally witnessed. The one you see could be considerably and wondrously different in several ways.

In this description, I'm going to limit us to what can be witnessed with just the unaided eye. Sight 19 is devoted to observing total solar eclipses with binoculars and telescopes.

The Approach of Totality

In the final minutes before totality, the pace of change quickens, and eerie alterations in the sky, the air, and the landscape bring a sense of marvelous foreboding.

The *temperature drop* may be as great as 20° F or more if the eclipse is occurring in the middle of a clear warm day, and a resultant *eclipse wind* may blow in just before or at totality.

Birds and animals may exhibit strange behavior, with the birds flocking confusedly to their roosts.

Most important, not just the intensity but the *color and quality of sunlight* changes, coming as it does now entirely from the edge of the Sun. Such light, shining more grazingly through a greater amount of solar gas, is somewhat dulled and reddened but makes the landscape look oddly drained of color, have a dim yellowish cast, or take on a dull silvery-straw hue.

Reflections of the Sun in water and glass throughout the landscape go from their usual strong, widespread glitter to mere isolated gleams.

Shadows become bent, those of fingers looking like claws, and the dappling of sunlight under leafy trees turns to multitudes of *thin crescent sun-images*—true pinhole pictures of the Sun produced by chinks in the foliage.

When is a shadow not a shadow? When it is part of *shadow bands*, eerie strips of duskiness that may begin to swim across the entire landscape (especially visible on sand, snow, and other white surfaces) a few minutes before totality (and after) and that move with the same direction and speed on the ground as the prevailing wind does high above the surface. What are the dark stripes if not shadows? They are the dimmer areas in the variations of light produced when the crescent (not quite fully covered) Sun becomes so thin that it is twinkling in the turbulence of our atmosphere.

The planet *Venus* often becomes plainly visible in the general vicinity of the Sun by a few minutes before totality.

But now there may be seen the final and most truly awesome prelude to totality itself: the *approach of the Moon's shadow.* This central or "umbral" shadow of the Moon is made more readily apparent by the presence of a little haze or high, thin clouds. It looks something like the purple or blue-black mass of a mighty thunderstorm rising up, but this "storm" is typically 100 to 200 miles wide and approaching at a speed of over 1,000 miles per hour. As it nears, the vast shadow's rate of advance up the sky becomes faster and faster. Will totality begin when the shadow has swept across the entire sky? No. It will begin when the shadow's forward edge reaches the Sun.

As the shadow's edge nears the Sun, a brief—very brief—phenomenon to look for in the instant of going into or out of totality is the *chromosphere* ("zone of color"). This is an only 600-mile-thick zone of gases just above the surface of the 860,000-mile-wide Sun that flashes out pink. You may not catch it going into totality, but there is another phenomenon you might experience. Suddenly, the glare of the remaining solar crescent, until now still dazzling and dangerous to even glance at, breaks into a string of fiery dots called *Baily's Beads.* These are the last (or, at totality's end, the first) sections of the solar crescent seen through lowlands on the edge of the Moon.

The Total Eclipse

Totality! Full darkness. The pearly *corona* of the Sun instantly leaps into view, a gently glowing wreath around the jet-black, profoundly round silhouette of the Moon, which is now completely hiding the literally blinding ball defined by the Sun's so-called surface or "photosphere." Only now, during totality, is it safe to stare at the Sun without special eye protection, for the light of the corona is only about as strong as that of the Full Moon.

But you the observer do not feel that the landscape is lit as brightly as by a Full Moon or twilight sky. The fall to darkness was so precipitous in those last few seconds before totality that your eyes couldn't completely adjust and felt more like they were suddenly plunged into deep night. It was as if someone rotated a dial to dim the light to darkness in a room in the course of a few seconds—except that the room was all of outdoors, for up to a hundred miles and more around you. It sends a chill of wonder down the spine.

Yet the corona is not the only light shining. For all around you on your horizon is a wide, glowing band of orange or red, the *360° sunset.* This is light leaking to you from lands beyond the Moon's central shadow, coming from so many dozens of miles away that it is ruddied by its long passage through the atmosphere. Most of the sky is a deep midnight blue, in which careful observers can usually detect a few bright planets and stars glimmering (if you wear dark sunglasses or goggles up until totality, you may see many more stars).

But the centerpiece of totality is what keeps drawing your gaze back: the dark moon surrounded by pearly corona, a sight so awesome that some eclipse-goers have called it "the Eye of God." The corona is the outer gases of the Sun, stupendously hot but so tenuous that it shines with a safe, gentle radiance—one only visible naturally during a total eclipse of the Sun. The corona is complexly structured, and some of this exquisite form can be noted by the unaided eye. The overall shape of the bright part of the corona looks like a dahlia blossom in years near "solar maximum" (see Sights 5 and 38) and more irregular, with long "petals" here and there, in years near solar minimum. The naked eye can sometimes follow faint outer streamers of the corona extending as far as several million miles (several apparent solar diameters) away from the Sun. And in the innermost corona, right at the dark edge of the Moon, you may be able to glimpse a startling sight—a few tiny but vividly red plumes. These and all the exquisite details of the corona are really properly seen with binoculars or telescope (see Sight 19). But if you do observe any of the little ruddy tufts with your unaided vision, know that you are seeing *solar prominences*—fountainlike structures above the solar surface that are each large enough to easily dwarf the Earth.

Totality lasts only a few minutes, never more than seven. As its finish nears, the sky is brighter in the west—the rear edge of the shadow is approaching the Sun. Then suddenly, at totality's end (sometimes at its start, too), another phenomenon may be observed. It is the single most staggering sight that a skywatcher is ever likely to behold: the *diamond ring.* Smaller and more precise than a Baily Bead, and appearing like a mighty star, the light from a single point of the Sun's blazing surface bursts through the deepest valley on the edge of the Moon. It is a "diamond" on the still visible ring of the innermost, brightest corona. But this is a diamond brighter than all those of Earth combined. Initially far brighter than the Full Moon yet concentrated for a few soul-shattering seconds into a point of light, it brightens and brightens further—then turns into Baily's Beads or a wire-thin blinding arc of exposed Sun. Eyes must turn away; the light of common day is rushing back in. Totality is over.

A tape recorder running during a total eclipse of the Sun picks up even the most meticulously prepared and normally impassive people shouting, crying, and babbling expressions such as "Oh my God!" and "Wow!" over and over again. But that is not surprising. For totality is a time, and a state of heaven and Earth, like no other. In fact, it feels like a time beyond time, as if the reality we learned to construct as small children had been ripped open to let burst through a gleam of what is eternal, beyond words or even thoughts, a blast of raw and pure wonder.

Be Sure to See One

You will almost certainly have to travel to see a total solar eclipse in your lifetime, and quite possibly the journey will be far, to another country. Before the 1970s, people had to make all the plans and arrangements for an eclipse expedition on their own. Fortunately, by the 1990s eclipse vacations had become a full-fledged industry for a number of tour agencies, and this continues to be the case. Most people's best bet will be to look through the pages of the popular astronomy magazines in the year and months before an eclipse to find the ads of such agencies offering eclipse vacations.

The good news for you if you live in the United States and are reading this book before 2017 is that there will be total solar eclipses passing over parts of your nation on August 21, 2017, and April 8, 2024. As a matter of fact, a region that includes southeastern Missouri, southern Illinois, and western Kentucky experiences totality on both occasions—two total eclipses of the Sun in less than seven years! Although the average interval between total solar eclipses at a given location is about 375 years, waits of just a few years between these events do occur. By contrast, a place will sometimes have to wait over a thousand years (Jerusalem has no total solar eclipse between 1133 and 2241). Even a huge country may go decades without being touched by totality (there is no total solar eclipse anywhere in the continental United States in the thirty-eight years after 1979).

Appendix A gives the statistics and locations for all total solar eclipses in the world between 2008 and 2024. What's important is that you do what it takes to get to at least one of these events. A heavy overcast sky can hide all the glory. But even if you find yourself under rather low, moderately thick clouds as totality begins, there is still a sight that may make you weak in the knees with awe. Under such circumstances, the 1,000-plus miles per hour approaching Moon shadow is not seen until the last second. And then its arrival is like a shutter instantly closing over the whole sky.

Meteor Shower or Storm

They are as much as several hundred times faster than jet planes. They come in endlessly different levels of brightness, paths, colors, glow-trail durations, burstings, and other behaviors. Sometimes they are brighter than

A Leonid meteor shower falling through the constellation Orion.

Venus and, rarely, may even rival or surpass the Moon itself, casting shadows that race across the ground in counterpoint to their own flight across the sky.

What are they?

They are meteors, popularly known as "shooting stars" or "falling stars." They are the usually fraction-of-a-second streaks of light seen when a piece of space rock enters Earth's upper atmosphere and burns up from the friction of its stupendous speed—a velocity of up to about 140,000 miles per hour in the case of meteors that encounter Earth head-on near dawn.

Meteors are so surprising and impressive that they almost always elicit a gasp or exclamation from their observers. Most people in today's indoor, head-down world of widespread city light pollution have probably only seen one or two falling stars in their lives. Little do they know that on the same nights each year—the dates when Earth is passing near the dust-strewn orbits of certain comets—as many as dozens of falling stars per hour may be seen. Such a display is called a meteor shower.

Witnessing a Meteor Shower

A **meteor shower** is a significant or noticeable number of meteors all appearing to diverge from a single spot in the heavens. Let's imagine what it would be like if you and a friend observed one of the year's best meteor showers.

The two of you walk out late on a clear August night to a large field behind your friend's house. The sky is dark and richly starred. You sit down in lawn chairs and look up at the majestic stillness of the stellar crowds.

Suddenly a moving light breaks the tranquility, spinning your head to one side to follow it as both you and your friend shout out.

You've seen your first meteor of the night. Its flight lasted less than the duration of a heartbeat—even less than the rapid heartbeat that is now pounding in your chest. Silence and stillness have resumed, but you are still stirred by what you saw. Then two minutes later—streak! Another meteor splits the sky. This one seemed to leave a glowing seam in the heavens for a full second or two after the meteor itself vanished. Your hushed but excited chatter about it with your friend is still going on when suddenly a third shooting star, this one vivid green, passes by, sputtering, flaring, finally bursting—you have seen a *bolide*, an exploding meteor.

These three meteors appeared in different parts of the sky. But if you traced their paths back from where they first became luminous, you would find the paths pointed to the same spot among the constellations, the **radiant** of a meteor shower. On this August night, the radiant is in the constellation Perseus, and you know that these meteors were all members of the Perseid ("offspring of Perseus") meteor shower. A fourth meteor, golden and slower than the others, now glides out of the south. That's almost the opposite direction from Perseus at this hour, so you suspect this shooting star may have been one from the declining phase of the Delta Aquarid meteor shower—a shower that peaks in late July and whose radiant is located near the star Delta Aquarii.

Why do the meteors in a shower seem to all emanate from a single point or small area of the heavens? The bits of space rock or dust (called **meteoroids** when out in space) follow more or less parallel paths in the vicinity of their parent comet's orbit and enter Earth's atmosphere basically parallel to one another. Why then do the meteors they produce seem to diverge? It is simply an effect of perspective visible whenever we see parallel objects coming to us from a distance. Meteors shoot out from a shower radiant for the same reason railroad rails appear to diverge from a point in the distance and snowflakes appear to diverge from a spot ahead of us when we drive a car in a snowstorm.

You and your friend see a "star" suddenly appear in the darkness, flare in brightness, and then vanish. It was very near the Perseid radiant. This *point meteor* must have been a Perseid meteor headed right at you! No need to fear, though. Even a meteor as bright as one of the brightest stars is usually caused by a speck of dust so small you could hardly see it if it were lying in the palm of your hand. Although that speck is so fast it would hit a spacecraft with the power of a hand grenade explosion, it usually becomes luminous roughly 80 or 70 miles high and is small enough to burn up completely by the time it is no lower than about 50 miles high in our atmosphere. Only one meteor in many millions manages to survive all the way to Earth's surface to be found as a **meteorite**. Furthermore, the meteors that do survive may all be tougher (and sometimes larger) bits of asteroids, whereas shower meteors are believed to be all, or almost all, more fragile particles derived from comets.

A dim meteor now races straight down the eastern sky. This one came from neither the Perseid radiant nor the Delta Aquarid radiant. It was likely a **sporadic meteor**, which is a meteor derived from no known shower. Every night of the year, even when there is no important meteor shower, there is at least a smattering of these sporadics.

Several cold, meteor-less minutes pass. Even during one of the best showers, such as the Perseids, meteor watching requires great patience and perseverance. There is peacefulness in sitting out under the stars, however. Suddenly, your peace is shattered. The night is seared—a meteor brighter than even the planet Venus, therefore a **fireball**, rips across the heavens. In the midst of your shouts, you notice your friend's face plainly lit up, and he notices your shadow cast.

This bright fireball ends up being the single best highlight of the night. But the thrills are just getting started. During the course of the night, you see blue, green, yellow, orange, and red meteors, even a pink one. You see fast Perseids with long paths and slower ones with short paths. You see two Perseids at once, one meteor that seems to have a crooked path, several bolides, another meteor that leaves a visible trail for 10 seconds to the naked eye and 30 seconds in binoculars. Under your excellent sky conditions, late at night when the Perseid radiant is highest, you and your friend end up seeing sixty Perseids and almost twenty other meteors in your best hour of watching.

Meteor rates of one or more shooting stars a minute usually seem to satisfy even the most impatient watcher, and those of us who have a special passion for astronomy are filled to the brim with excitement and delight when we experience such a display. And yet while enjoying a strong meteor shower you can always dream of stronger. In one night in 1999 in Europe and the Middle East, and in 2001 in North America and the Far East, the Leonid meteor shower became a **meteor storm**—in each of these cases some observers saw rates of at least 1,000 meteors per hour for a while.

How to See and Record Meteor Showers

Use appendix B to determine when to look for the year's major meteor showers and where their radiants lie. If the radiant is only a third of the way up the sky (30°, three fist-widths at arm's length, above the horizon), only half as many meteors will be seen. If the radiant is overhead and the Full Moon is well up in the sky, you live in a big city, or there is a light cloud or thick haze all over your sky, you will see only about 10 percent of the possible meteors. More than about five days past New Moon or more than about five days before New Moon, moonlight is strong enough to make at least a serious reduction in the number of meteors.

Once you know what date and time of the night to expect a lot of meteors, there remain the questions of where to look in the sky to see the most meteors and how to keep an accurate count of the meteors you see. On the one hand, if you just want to enjoy a meteor shower casually and do only a rough count or none at all, then you only need to dress extra warmly, seek an observing site where much of the sky is visible, and look anywhere in the sky you wish.

On the other hand, if you want to try to make meteor observations you can submit for scientific purposes, there are several rules to follow. The greatest number of meteors will be seen if you watch rather high in the sky and about 20° to 40° away from the shower's radiant. Your *effective field of view* is a circle of about 50° radius (about 98 percent of all meteors observed are within this circle). If more than 20 percent of this field of view becomes filled with clouds, you should stop your official count. Also remember that each individual must keep his or her own count. Do not include a meteor in your tally if you would not have seen it without getting a warning shout from a fellow observer. Try to observe and count in intervals of 15 minutes or less. Keep separate counts of meteors from each shower currently visible and of sporadic meteors. If you can, add notes about the brightness, position, trail, and other characteristics of each meteor. The best way to try to keep track of all this information is by speaking it into a tape recorder.

A Few Notes on the Best Showers

Let's conclude this chapter with a few fascinating facts about some of the best showers.

1. The *Perseid meteor shower* (also known as "the Tears of St. Lawrence") is derived from what has been called "the single most dangerous object known to humankind." That is Comet Swift-Tuttle, which last passed Earth in 1992 and will not be back again until 2126. The solid nucleus of this comet is possibly almost 20 miles wide, making it perhaps the largest object we know that in its current orbit could at some distant future date collide with Earth. If it did collide, the calamity could be much greater than the one that destroyed the dinosaurs and about 75 percent of species on Earth 65 million years ago.
2. The *Geminid meteor shower* may be the only major shower derived from an asteroid—unless the Phaethon asteroid is really an extinct comet nucleus.
3. The *Lyrid meteor shower* is the earliest recorded (it was noted back in 687 B.C.).
4. The *Quadrantid meteor shower* is normally the most sharply peaked of the major showers (but its maximum numbers of 40 meteors or even more than 100 per hour are only seen if the radiant is high in your sky in the dark for the hour or two of the peak).
5. *The Orionid and Eta Aquaird meteor showers* are debris from the most famous of comets, Halley.
6. The *Leonid meteor shower* is capable of producing peaks of hundreds or even thousands of meteors per hour on certain nights in years around the time of their parent comet's passage. (Comet Tempel-Tuttle passes Earth about every thirty-three years and last did so in 1998.) In 1833, a

Leonid meteor storm of possibly up to 60,000 meteors per hour woke people out of their beds in Boston and became known as "the Night the Stars Fell on Alabama" down South. In 1966, a brief Leonid storm peaked over the U.S. Southwest at rates that may have reached as high as 500,000 meteors per hour in the best minute!

Sight 4 Fireball Meteor

There is probably no experience in astronomy more gripping—more literally knuckle-whitening—than seeing a bright fireball meteor. When the diamond appears at a total solar eclipse, it is more awesome, so awesome it is beyond gripping (you and your body forget what to do and afterward forget what you did). Moreover, the diamond is a great soaring peak of experience at the summit of the mountain of wonder that the eclipse has already put you on (this could also be said about some moment of supreme beauty in the midst of a great Northern Lights display). But when a fireball meteor occurs, you usually have no warning that anything spectacular is about to happen.

In fact, there is often no warning that anything at all is about to happen. One moment, you are gazing peacefully at the stars, or perhaps just at something terrestrial in your field of view. The next moment, bright light assails you and your unbelieving eyes follow an impossibly intense, fiery, and sometimes vividly colored mass of light as it rips a gash of radiance across the night sky. A second later, darkness returns—unless a glowing trail from the meteor burns on, fading slowly from the sky, but never fading from your memory.

Minimum Fireballs and Shadowcasters

Actually, a meteor doesn't have to be quite bright enough to light up the night to be classified as a fireball. As we saw in the previous chapter, the definition of a **fireball** is any meteor that shines brighter than the brightest planet. That would normally be Venus, but since Venus is not visible as often as Jupiter, the lower limit of fireball brightness is sometimes considered to be

Jupiter, the second brightest of the planets. A minimum-brightness fireball will not necessarily make you dig your nails into your palms or shout out loud, but it is still a marvelous sight. Consider that it is the brightest point (or, if bigger, brightest intense concentration) of light in the heavens, not just moving rapidly but also usually varying radically in brightness, color, and shape of its sizable head as it moves, either trailing sparks or leaving a luminous trail of some sort.

Now ratchet up the brilliance a few major notches from this minimum example. I think of some of my most memorable fireballs. Two of them were just the best of many in an incredible display of all-bright Leonid meteors in the only clear (and moonless) hour one night in 1998. I'm pretty sure that was the night I first started using the term *shadowcasters*. The Leonids are among the fastest of meteors, and the brightest one that night not only lit the ground but it also flared with frighteningly fluctuating and flashing light as it flew high over me on the remote stretch of country road where I stood. I am not speaking exaggeratedly or metaphorically when I say it was flashing like lightning. It actually was, on a November night of tranquil weather, like suddenly experiencing moderately close multiple-pulse lightning. It was so startling that I couldn't help but have a brief reaction of fear mixed in with my overwhelming and triumphant wonder and admiration. Not many minutes (but several fireballs) later, an especially bright fireball soared overhead—and went out within the pentagon of stars that is Auriga the Charioteer, leaving a glowing trail that seemed to burn on as if literally branded on the sky. I watched in utter amazement as winds high in the atmosphere rapidly distorted and evolved the slowly fading trail. In a minute or two, it had the appearance of a luminous oval smoke ring whose longest axis was about three times the diameter of the Moon. In another minute or two, it had the shape of a barbell. Then of a streak and two puffs of smoke. Some trace of the trail remained visible to my unaided eye for a full eight minutes. I wish I'd had my binoculars with me to see intricate details in it and watch it shine for many minutes more.

Only a few times in your life might you be lucky enough to see several shadowcaster fireballs in the same night during a meteor storm or unusually bright meteor shower. Venus-bright fireballs can be seen in many years, even most years, if you do a lot of watching of major meteor showers. A few of the showers are more likely to produce fireballs (the most famous in this respect are the slow, though not very plentiful Taurids of late October and early November nights). But most of the brightest fireballs will come at a time unconnected to meteor showers, because they are caused by larger, tougher pieces of asteroidal material, not the fragile cometary particles that are found in most meteor showers.

Electrophonic Fireballs

Sometimes an atmospheric entry that is more grazing will allow a meteor to last longer—as will the slower entry that occurs in early evening when meteors are having to catch up to the sum of Earth's orbital velocity and rotational speed. A case in point was the altogether most exceptional meteor I've ever seen: a majestically slow, 10.5-second duration, Half Moon–bright, electrophonic and probably meteorite-dropping fireball. Its head was a mass of green, fiercely writhing flame slowly emitting lingering and down-arcing fragments that themselves rivaled the brightest planets. Back in 1982 when it occurred, I collected dozens of eyewitness reports and enlisted the help of David Meisel, the president of the American Meteor Society. We were able to rule out a satellite reentry and calculate the likely size and mass of the original object and of the pieces it probably dropped a mere few miles out to sea. Meisel was also able to determine a remarkable orbit that the object had been pursuing around the Sun.

In my book *The Starry Room*, you can read a twenty-two-page-long chapter devoted to this 1982 fireball, including some discussion of another remarkable aspect of it: there were not just eyewitnesses but also ear witnesses of this object. Furthermore, the sound it produced was not the minute-or-two-after-the-light sonic boom that some low fireballs produce. It was a hissing or swishing sound experienced simultaneous with the light of the meteor. The meteor was probably electrophonic—that is, producing sound by electromagnetic radiation. In the case of fireballs, the radiation is believed to be very-low-frequency radio waves caused by the object's disruption of Earth's local magnetic field. Some researchers believe that these radio waves can produce vibration in parts of the landscape near the observer, and the hissing is that vibration. But it's even possible that the radio waves directly stimulate the nervous system in such a way as to simulate the sensation of sound. There are reports of people who were nearly deaf experiencing this anomalous, simultaneous sound from a fireball very plainly.

How common is it for even bright fireballs to produce simultaneous sound? Pretty rare. How common is it for a fireball to produce meteorites? Very rare. And there are only a few cases of people's observations of a fireball leading to the finding of meteorites from it (again, see my book *The Starry Room*). Yet, even a remote possibility of finding these ancient mementoes of the early solar system newly arrived from outer space in a fresh state is important. It makes even more compelling the idea of learning what to look for and note in the frantic second or seconds that one of these prodigal objects accosts your senses.

Tips for Would-Be Watchers

As stated earlier, you can increase your chance of seeing a fireball by watching the major meteor showers and, slightly, by keeping a special eye out in the early evening. One additional specific and one additional general piece of advice can be given.

First, the specific suggestion: watch for earthgrazers when there is the greatest chance of seeing them—around the time the radiant of a major meteor shower is rising. Earthgrazers are meteors that encounter the atmosphere at a shallow or even tangential angle and pursue long paths across the heavens. They are by no means always fireballs. But their long tracks increase the chance that you'll catch a sight of them. In the big Leonid meteor year of 2001, an observer I know was out watching around midnight, soon after the Leonid radiant had risen. He witnessed a Leonid earthgrazer cross the entire sky so slowly and majestically it was like a train—"the Midnight Express," he thought.

A second piece of advice for detecting fireballs is general and pretty obvious but nevertheless resoundingly true: go out and look at the night sky a lot.

There is another, and crucial, part to the topic of "how to see" a fireball. I refer to what you should remember to perceive and note in the brief and startling time that you do see a fireball. If you don't review and rehearse now what you'll do then, you will almost certainly fail to apprehend some of the most important properties of the fireball.

You should, of course, try to notice where the fireball passed in the sky—either in comparison to background stars or to objects in the landscape. Sometimes a fireball is so bright that it entirely overwhelms the stars it is passing near, or even an entire section of the sky, or even the entire sky if it dazzles your vision. Perhaps you can note the fireball's beginning, middle, and ending positions in relation to tall trees or buildings in the landscape. If you can, make certain you mark or memorize the exact spot from which you were watching. Try to determine not just the cardinal directions but the more precise azimuths the object was in. Angular altitude in the sky can be deceptive (we tend to feel that anything above about 45° high is virtually at the zenith). How bright was the fireball? Ask yourself if it was really brighter than Venus or Jupiter (if these planets are present, your task will be easier). Ask whether it really rivaled the brightness of a thick crescent or even Half Moon (did you see a fellow observer lit up or shadows cast by it?). How long do you think the fireball was visible? This may be hard to guess and novices often overestimate the duration (remember, a second is long enough to count at normal speaking speed "One thousand and one . . ."). Note any color, details of the trail, and whether the object fades out or disappears with a terminal burst. And, of course, be sure to note if you hear any sound associated with the fireball: either a hiss or a swish as it passes or a rumble a minute or two after it has vanished.

Sight 5 THE NORTHERN LIGHTS, OR AURORA

There is no doubt what is the heavens' greatest "light show."

Few sights in the night sky offer to the eye strong color. Likewise, few display rapid motion. The subjects of our previous three chapters (total solar eclipses, meteor showers, and fireball meteors) can offer color and at least brief outbursts of fast motion and/or sudden change of appearance. But there is one sky phenomenon that features sometimes intense colors and innumerable intricate structures that not only change dramatically in the course of seconds (or even fractions of seconds) but also move with every kind of motion imaginable: pulsing, rotating, projecting, rippling, waving, dancing, whipping, radiating, flickering.

Partly because prominent displays of it are only common at fairly high latitudes, but largely because it is truly as eerie and beautiful as anything in existence, this very real phenomenon has a legendary status. Even its names are magical. For I am speaking, of course, of the **aurora borealis**, better known as the **Northern Lights**.

Progression of Majesty

The phenomenon called the Northern Lights in the northern hemisphere has its counterpart in the southern hemisphere, where it is called the "aurora australis." The basic name that scientists favor wherever the phenomenon occurs is **aurora**.

Before we concern ourselves with what causes the aurora and when to see it, let's consider the kind of appearance it offers.

A typical display at midnorthern latitudes (between about 30° N and 50° N) may begin with an amorphous glow low in the north. It may be hard to distinguish at first from the skyglow of city light pollution in the distance. But when changes in the glow's brightness start becoming noticeable over the course of a few minutes or less, you should get suspicious. If the light then begins to get arranged in one or more horizontal *arcs*, you know that an auroral display is under way. These arcs may form, disappear, then get more prominent, and they may fluctuate in brightness. The question is: How much stronger and more developed will the display get? If you're lucky, the next stage is often the formation of vertical *rays* in, or sprouting up from, the arcs. Some color, most often a pale green, may start to become visible. This is when things begin to get thrilling. Here and there, a ray may shoot up, perhaps towering most of the way up the northern sky. Greens may intensify, and other colors—most commonly red—may appear. New large patches of glow can break out higher or here and there in other directions—northeast and northwest, east or west, even southeast or southwest. The vertical rays may form auroral *curtains*, which can even simulate the rippling of linen curtains being blown by a wind. Finally, with colors and forms pulsing in too many places and ways for your delighted attention to follow, curtains may come together to form high up near the zenith a spectacular and usually colorful crown of radiating rays—an *auroral corona.*

Views for Southerners and Northerners

An auroral corona is what you see when you stare almost straight up into an auroral curtain. But it is not common to get bands of aurora far enough south for a curtain to come over a person as far south as say 40° N. At least when Northern Lights do appear high in the north at such a latitude, they are usually more or less vivid red, because they are produced by fluorescence of hydrogen (and some oxygen) at a high altitude in the atmosphere (up to as high as about 500 miles above ground level). The green light is caused by excitation of oxygen only, lower in the atmosphere—at altitudes of as little as about 50 miles. Satellite photos show that the zone of auroral activity encircles the poles of Earth—the *magnetic poles*, offset by hundreds of miles from Earth's geographic poles—like a necklace (though not detectable in visible light in the hemisphere of Earth that is sunlit). Within this pole-encircling *auroral oval*, bands of aurora occur. Under favorable geomagnetic conditions, the oval intensifies and expands, moving south. If it reaches the latitude of, say, northern New England, observers there may be able to stare up into an auroral corona and see patches of aurora break out south of them. A few hundred miles farther south—say, in Long Island, outside New York City—observers would be seeing the red top of the hundreds-of-miles-tall band only a fraction of the way up their northern sky.

Don't be discouraged if you live as far south as 40° N, 35° N, or even 30° N: there will still be occasions when you can glimpse some lovely auroras. And nowadays there are ways you can be notified when an extreme south-reaching display is about to occur or is occurring. I live just south of 40° N, in southern New Jersey. It's not every year that I get to see an impressive aurora from here. But in years when I do, there are usually several displays. And marvels do occur: vast patches of red can burn high, sometimes bright enough to be visible right up next to the Moon, then spread like instant wildfire to another section of the sky; first delicate and then powerful rays of several colors can shoot tall, restlessly reaching for the zenith. In one memorable case, I saw the sky through the trees low to my north become incredibly, intensely green and then even brighter (though whiter) so that it was as if a light-wasteful city of 500,000 people had suddenly sprang up a few miles north of me. It was a horrible thought that if such a city did someday appear (where there are now a few dozen miles of mostly environmentally protected forest), such light pollution would assail my sky and compromise all the stars in it. But dread gave way to joyous wonder when I reminded myself that this was not that nightmare scenario (that even humankind's most thoughtless and wasteful inclinations might never be allowed to produce, at least not for decades). It was the stunning natural equal in amount of radiance (actually much stronger in true luminosity, considering the light was from auroras hundreds of miles away!). But it was an uncommon and brief natural display I was seeing, one that I knew would soon subside back into restful, star-clad darkness—which it did.

What greater wonders than these await observers of the Northern Lights at higher latitudes? My own very limited experience stems mostly from a few marvelous weeks I spent in northern North Dakota one especially aurora-active summer. My notes from memories of those displays would fill many pages. Here I can only hint at the staggering beauties I experienced. I saw whole-sky coverings, in three major and two minor colors, in pulsations, flickerings, propagations, clouds, strandwhippings, rays, multiple arcs, curtains, coronas, and landscape-illuminating, shadow-casting patches of brilliance. But the aurora can get so bright as to hide the brightest stars even for viewers at 40° N and farther south. One thing that especially sets displays at northern sites apart from more southerly ones is the amount and kinds of motion—the aurora is often phenomenally more active, varied, and swift in its movements at northerly sites. In North Dakota I saw not just green, red, purple, yellow, and orange in horizontal arcs, vertical rays, and curtains of rays. The curtains rippled. The aurora flashed with glints, flamings, flingings, and dashings. I saw an auroral form that seemed to show three-dimensional spiral motion like a vast tornado of light. Almost a quarter of the heavens filled from low to high with what seemed to be a vast tree of aurora whose branches were growing still (motionless) as the spectacle slowly began to fade. I saw auroral angels reminiscent of the spirits flying around in the climactic scene of the movie *Raiders of the Lost Ark.* And I saw the full development of a sky-encircling, rotating, many-armed color organ—the auroral corona.

Cause, and Tips for Seeing

The aurora is a complex phenomenon and is still far from perfectly understood. Basically, it starts with intensifications in that ceaseless stream of atomic particles from the Sun that is called the *solar wind.* Normally, Earth's magnetic field shields us thoroughly from the solar wind. But when a *coronal hole* in the Sun's outer atmosphere leaks particles, or when bursts of them are spewed out at up to millions of miles per hour by events on the Sun called *solar flares* and *coronal mass ejections* (CMEs), then our planetary magnetic field is overwhelmed and the particles are accelerated with great energy into our atmosphere in the pole-encircling auroral ovals (around both the north and south magnetic poles). The upper atmosphere gases are excited and fluoresce, acting as visible tracers of the lines of magnetic field that may drastically vary their positions and strengths across hundreds of miles in mere moments.

The solar activity that produces *geomagnetic storms* and sometimes auroras varies in a period of very roughly eleven years from one *solar maximum* to the next. So auroral displays tend to be more frequent and strong in the years around (and perhaps after) solar maximum.

Solar minimum is occurring in 2006–2007, with the next solar maximum

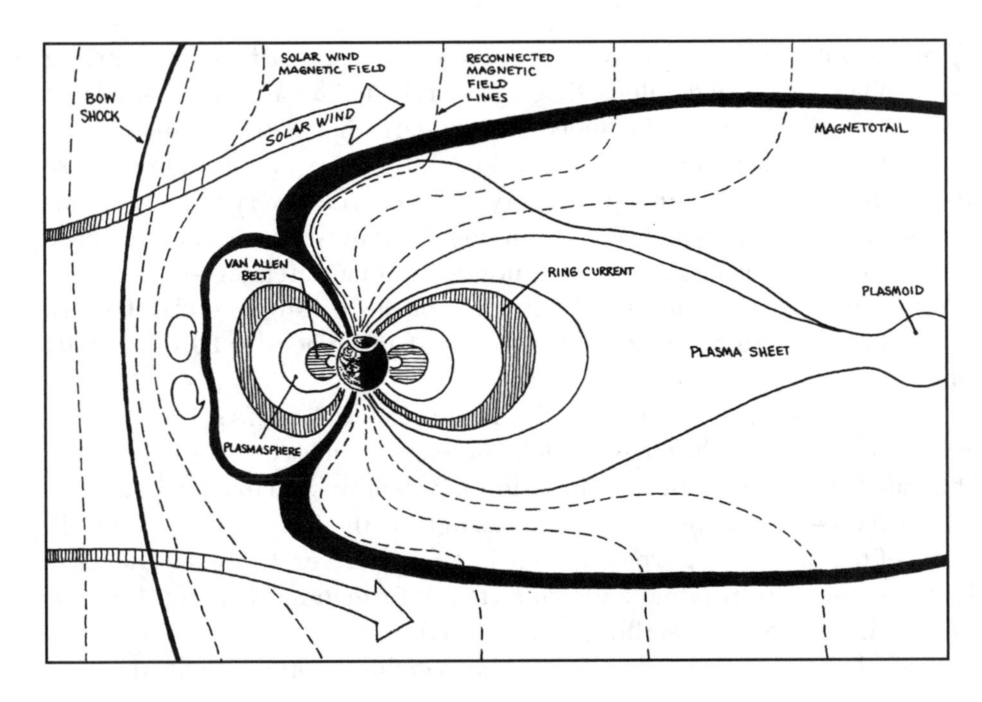

Structures of Earth's magnetic field related to the production of the aurora.

probably not coming until 2010 or 2011. But there are always a few major outbreaks of auroras even in the years around solar minimum. And with twenty-first-century science and communication, there are some dependable ways to ensure that you will catch these outbreaks. An outstanding starting point for those who want to keep track of solar activity and aurora possibilities is www.spaceweather.com. This is also one of the best all-around astronomical Web sites, with information on upcoming and recent sky events of all sorts and access to wonderful collections of images of those events. If you check Spaceweather.com frequently, you will always find interesting news, data, and pictures there, and you are unlikely to be caught unaware by a great auroral outbreak. To be really safe, though, you can subscribe to the *Sky & Telescope* AstroAlert service, guaranteeing that you'll get e-mails notifying you when conditions are good for auroras. Or you can even receive phone calls from Spaceweather.com notifying you that an aurora is in progress where you live! For more information on these services and other resources related to the Northern Lights, see the sources section at the end of this book.

Bright Satellite or Spacecraft

I must confess that I've never been one of the biggest fans of observing satellites—artificial satellites, that is (natural satellites are *moons*—rocky or icy bodies orbiting around planets). My primary objection to artificial satellites is that they *are* artificial: there is no deep mystery about them because they were designed by humans for very specific, usually very utilitarian purposes. Even worse, much of what is orbiting Earth must be considered "space junk"—booster rockets, debris, and nonoperational satellites—even by the most charitable of definitions. At the time of this writing, the U.S. Department of Defense was tracking approximately 10,000 objects in Earth orbit, 95 percent of which was junk.

Now having expressed my negative feelings about satellites' lack of natural mystery and the clutter their debris creates, I must immediately admit that satellites also have their positive side and their undeniable fascination to all observers—including myself, increasingly as the years go by. After all, some of the utilitarian purposes of satellites are inherently noble. They may be performing science that adds important information to various fields of knowledge. Weather satellites track hurricanes and other dangerous weather, helping to save many lives. Other satellites monitor environmental

Iridium satellite flare.

conditions, acting as tools to help us protect our biosphere from the destructive aspects of many kinds of short-sighted human activities. Then there are satellites whose images are both information and art. Who would not wish to see the Hubble Space Telescope, whose own eye has revealed to us so many new aspects of our universe, often in the form of pictures worthy of being hung in art galleries?

Finally, there are orbiting artificial objects that contain humans. These (and robot probes that leave Earth orbit, often to visit other planets) are usually called spacecraft. And even if there were no practical benefits derived from sending men and women into space, there would remain the inquiry and adventure, the new experience and fresh perspectives that these missions provide. Here on Earth, there are few thrills more potent than that of watching a point of light glide across the night sky with you knowing that there are humans in it looking down over large regions of the world.

The Sight

I was born just before the space age began and became interested in astronomy as a young child. So I actually have some vivid memories of seeing satellites and rockets throughout my life. But the ones most relevant to our interest here are those that date from the 1980s to the present. The first of these was the space shuttle.

Most U.S. space launches have been from Cape Canaveral, Florida, into fairly low orbits inclined at 28° to Earth's equator. This guarantees that they can never be seen from much farther north than 28° N (the latitude of Cape Canaveral) or much farther south than 28° S. In the 1980s, it was rare for the space shuttle to have a mission that took it far enough north to be seen where I lived (and still live), near 40° N on the East Coast. But I remember vividly and fondly seeing the shuttle from New Jersey for the first time ever on one of its exceptional flights into a more highly inclined orbit. I and my mother, who was always a fan of space exploration, watched on television the launch from Florida during the supper-hour world news. We knew that in 2 or 3 minutes the shuttle would already reach the latitude of New Jersey and that one of its rocket stages would still be firing as it passed us, fairly low in the southeast and east, many miles out over the Atlantic. We briskly moved outside and after only a slight and suspenseful delay—there it came! It looked like a very bright star, due to the flame of the rocket engines, a shape of flame I glimpsed in my binoculars, even in twilight. I tried hard, but I wasn't quite able to get my mother a view through the binoculars. No matter, she thrilled to the naked-eye sight with me, and we savored this experience together. We watched as this human-bearing star hastened away, suddenly dimming out of view when the rocket was spent.

My visually best view of the shuttle heading up the East Coast after a launch came years later when I watched for it in the lonely cold at one of my favorite sites, a field near a lovely local pond. This time the launch was in the midnight hour, and I started wondering if there had been an extensive delay, for I was not seeing it. Then suddenly here it came from low in the south or southeast, becoming surprisingly prominent. It passed not far from Mars, which at that time was impressively bright, but the shuttle's flame outshined the planet and presented the naked eye with a color similar to that of the orange-gold planet. This time I was able to catch and follow it for a little while in a spotting scope that was more powerful and light-gathering than binoculars, and the view of the orange crescent- or banana-shaped flame was spectacular.

In the 1990s, there started to be a reason for the space shuttle to pursue an inclined orbit that would bring it far enough north even to be seen from southern Canada: it was transporting parts up to build the International Space Station (ISS).

The ISS is in an orbit inclined 52° to Earth's equator, so it can be spotted at one time or another from most of the inhabited places in the world. As the ISS grew, it had more surface to reflect sunlight, and it brightened, rivaling and even surpassing the shuttle. I have managed to see the shuttle pursuing the ISS and the two joined together as one light, and I've finally even been lucky enough to catch them in the act of separating.

As I write this book, the hope is that the shuttles will fly for at least a few more years before being replaced. In the meantime, the ISS has grown to 240 feet across, which is big enough at its average altitude of about 250 miles for its shape to be glimpsed clearly in telescopes (catching the fast-moving object in a telescopic field of view long enough to perceive its shape is tricky, though). What's easy is seeing it with the naked eye when it passes fairly high in your sky. At such times, the ISS can outshine all but the brightest star. As a matter of fact, there are exceptional moments when its huge solar panels reflect more light in your direction and the point of light flares to rival Jupiter in brilliance. The ISS can be seen at passes that occur within a few hours after sunset or a few hours before sunrise. Most of us would rather be up in the evening. At such a time, it's always a thrill when you first catch sight of the ISS mounting your sky and brightening. But the predawn passes do offer the thrill of having the ISS swiftly swell into brightness from out of invisibility—you are seeing it at the moment when the astronauts are experiencing sunrise bursting suddenly into their windows.

How to See Satellites

Satellites shine only by reflected sunlight. But because most of them orbit a few hundred miles above ground level, they can still catch sunlight for a few hours after sunset or before ground-level sunrise. It is easier to launch an object into space in the direction of Earth's rotation, so most satellites move roughly west to east (though there are some in polar orbits—north to south or south to north). They are usually easy to distinguish from meteors and airplanes. Meteors streak across the sky in a fraction of a second or at most a few seconds and often vary in appearance radically in the course of their brief flights, which sometimes includes a glowing trail behind them. High-flying aircraft may take many more seconds than meteors to cross the sky, but they have multiple and/or flashing lights on them and sound from them can often be heard at some point. Satellites typically require a few minutes to cross whatever part of the sky they remain visible in and then fade in a few moments when they reach the cone of Earth's shadow in the evening (or kindle into light when they escape that cone a few hours before sunrise comes to us at ground level).

If you can see many stars overhead, the clearness and darkness of your sky is probably good enough for you to behold ten or more satellites a night with your naked eye in the hour or two after dusk fades. How many you actually notice depends on how ably you scan for slow-moving points of light. But the way to make sure you catch any or all of the bright satellites, and to know which ones you are seeing, is to consult one of the satellite-tracking Web sites. The most popular is www.heavens-above.com. You can get information about many of the prominent naked-eye satellites for your location. The tables for each satellite give the time, altitude, and azimuth of where the satellite first appears, where it reaches its highest, and where it disappears. The brightness of the satellite when highest is given. In some cases, you can even bring up two maps for a satellite pass: a star map showing where the satellite will pass among the constellations as seen from your location, and a surface map showing the ground track of the satellite. The latter marks the locations on the surface of the Earth that the satellite is passing directly over. Thus, the ground track might show the ISS passing right over Pittsburgh but the table indicates that it will appear less than halfway up the western sky as seen from Philadelphia at that moment.

There is a spectacular kind of satellite display that you really do need the help of heavens-above.com to know when and where to see from your location: *iridium flares*. The almost 100 communications satellites of the Iridium series pursue highly inclined orbits—they nearly pass over both poles—and are visible from anywhere on Earth. Each of these satellites has extremely reflective antenna arrays. At certain times, the array is pointed just right to reflect a burst of sunlight to your location. When this happens, these otherwise dim satellites flare tremendously brighter. For no more than about 10 seconds you see the satellite appear, kindle to great brilliance, and then fade back out. Sometimes you will see an iridium flare that rivals the brightest stars or brightest planets. But at least a few times a month you should from your home location be able to see an even brighter iridium flare—up to a maximum brightness of magnitude –8. That is as brilliant as a thick crescent Moon and can cast a shadow at a dark site. The brightest iridium flares are easily bright enough to detect in broad daylight. Heavens-above.com calculates for you all daytime iridium flares that are brighter than magnitude –6.

THE MILKY WAY

On a summer evening, you find yourself far from city lights. The weather is unusually clear and cool for summer and the Moon won't rise, thin and elusive, until just before dawn. You've been outside a few minutes already, but under trees. Now you walk out into a clearing.

Naked-eye Milky Way in Sagittarius.

You stifle a cry of astonishment. It's not the seemingly countless stars sprinkled overhead that draws your amazement. It is a band of soft but prominent glow arching above you through the multitude of stars. This ribbon of radiance bridges the sky from northeast to south and exhibits extraordinary detail in places. In several places along the band, patches of glow are so bright you could believe they are a weather cloud lit by some nearby city—but there is no city nearby and the sky is cloudless. And yet they *are* clouds—clouds of stars thousands of light-years away along another arm of our spiral galaxy. In other places, there seem to be spooky bays and inlets of dark sky patterning the band—one dark tongue even splits the band in half, with one arm or branch of glow petering out and the other regaining strength to grow even more glorious, broad, and bright as it nears the southern horizon.

What is this band? Its slightly uneven course, its narrowing and broadening, its growing to a seemingly stronger flow as it gets down to the edge of sky and earth—all these features led some cultures to refer to it as the River of Stars. Other people regarded it as a road for the newly dead to travel to their dwelling place in eternity: the Path of Souls.

But the most famous name for this band that has come down to us from ancient times is a title that, even today, is known by almost everyone, even if they couldn't tell you what it is or what it looks like. The great sky-spanning band of dreamy radiance is known as the Milky Way.

From Cygnus to Sagittarius

I've already mentioned that the glow of this Milky Way band is produced by the combined light of vast numbers of faint stars—stars too distant to be seen individually with the naked eye. There is more to say in explanation of what the Milky Way is and why it is arranged in the form of a sky-spanning band—in fact, actually a heavens-encircling band, as we could see from midnorthern latitudes on Earth were it not for the fact that part of the Milky Way runs too far to the south for it ever to get above our southern horizon.

Before more explanation, however, let's just go out and examine the sight of the Milky Way.

Even before learning anything about what it is, people find the sight of the Milky Way one that stirs the heart and imagination. No other sight in nature seems so distant and vast yet so present, so dreamlike and yet so undeniably real in its fixed orientation to the stars and its intricate, varied details of internal structure.

Part of the Milky Way that can be glimpsed even in moderately light-polluted skies of large suburbs and small cities, or even on country nights of fairly strong summer haze, is the Cygnus Star Cloud. This patch of radiance is like a sort of glorious mist through which the stars that form the main pattern of Cygnus the Swan seem to be flying. One reason the Cygnus Star Cloud is the easiest part of the Milky Way to see under slightly adverse sky conditions is that it passes overhead for observers at midnorthern latitudes. Another reason is that it is truly one of the brightest regions of the band. Binoculars scanning through this region in dark skies can show more stars than perhaps any other place in the heavens. Telescopic fields here are star-crowded as well.

Just northwest of Deneb, Cygnus's brightest star, a vast cloud of interstellar dust intrudes almost all the way across the Milky Way band. But it is in Cygnus that the largest dark feature—the Great Rift—begins. This rift splits the River of Stars into two channels, one of which spreads and fades out to a trickle in Ophiuchus. The other, main channel runs down past Altair and in the little constellation Scutum, about midway up the sky, burns brightly in the roundish Scutum Star Cloud. This charming 5°-by-5° patch of radiance is often visible on somewhat hazy country nights when the lower grandeurs of the Milky Way are lost from having to shine through a longer pathway of absorbing and scattering humidity (water vapor).

In many northern industrialized countries, summer haze is greatly worsened by sulfur dioxide that is emitted in vast quantities from coal-fired power plants. But all Milky Way watchers long for those spells in summer when a strong cold front manages to sweep through and whisk away water vapor, dust, and air pollution. When this happens, a viewer far enough from city light pollution can see the dreamy grandeurs of the broadest and brightest section of the Milky Way band: the parts that flood over Sagittarius.

This region is rife with the tiny spots and patches of star clusters and nebulae bright enough to be glimpsed with either the naked eye or at least binoculars. M8, the Lagoon Nebula, is plainly visible to the naked eye in really dark skies, as are the two giant star clusters just above the stinger of Scorpius, M6 and M7. (These objects and others in this region are discussed in Sight 32.) The larger features of the Milky Way, however, are the Large Sagittarius Star Cloud and M24—also known as the Small Sagittarius Star Cloud. The latter is perhaps the brightest naked-eye knot in the Milky Way. It measures only about 2° by 1° and lies about 10° (one fist-width at arm's length) due north of the spout of the Teapot of Sagittarius. Closer to the northwestern part of the spout is what appears as a major puff of steam emitted from the imagined Teapot—the Large Sagittarius Star Cloud. When we look at this star cloud, we are staring in the direction of the glorious center of our galaxy.

Vision of Our Galaxy

The ancient application of the term *Milky Way* refers to the band of glow we have been describing. But in modern times we have decided that "the Milky Way" is also the name we will apply to the *galaxy* we live in. A galaxy is a collection of typically billions of stars, and the Milky Way is *our* galaxy. In other words, it is our home system of billions—perhaps a few hundred billion—stars, including the star we call the Sun.

What is the connection between the Milky Way Galaxy and the Milky Way band of glow we see? The band is the more densely populated equatorial disk of our pinwheel-shaped or lens-shaped galaxy. And it looks like a band in the sky because our own Sun, solar system, and selves are located within this equatorial disk. We see the disk from within.

One thing needs further explanation. Other sections of the Milky Way band can be seen in the evening in other seasons. The winter section of the

Milky Way arches high up the sky between Sirius and Procyon and across parts of Gemini, Taurus, and Auriga—but it is far dimmer than the summer Milky Way. The challenge with the winter Milky Way is to see it at all. If you behold it fairly plainly, you know you have a respectably dark sky. But why is the winter section of the Milky Way dimmer than the summer one? Because when we look among the winter constellations, we are staring within the plane of the galaxy's equatorial disk but in the direction of the outermost parts of our spiral galaxy. In summer, on the other hand, we are staring inward toward the center of the vast Milky Way Galaxy.

What we see in Sagittarius—the Large Sagittarius Star Cloud, the many nebulae and star clusters—is still only the outer part of the central glory. They shine mostly in the spiral arm of our galaxy that is the next one inward from our own. This arm is about 2½ to 5 times closer than the galactic center itself—a ball of billions of stars. There is so much interstellar dust in the 25,000 light-years or so between us and this central hub that its light never reaches us. Much of what we know about it comes from observations at other wavelengths of electromagnetic radiation—among them, radio and infrared wavelengths. But at least we can imagine what the central hub would look like without interstellar dust. It would be a big, eerie ball of glow that would light up its region of our sky perhaps as brightly as a crescent Moon.

What's more important is what we *can* see—if our own glaring lights don't blind us. When we look at the Milky Way band, we are staring at the largest structure of the universe that can be seen in detail with the naked eye. It is, one might say, the very face of our great galaxy—a face that a large fraction of the human race has now lost to the useless waste that is light pollution. You might also say that the dreamy grandeur of the Milky Way band, while fetching our spirits forth with its strangeness and the exotic allure of faraway places, is also an ultimately reassuring sight that stirs in us recognition: we are looking at our home.

The Milky Way is an extended glow. So to see even its brighter sections requires a respectably dark and clear sky. Some regions of its glow are dominated by stars closer and brighter than in other regions. Amazingly, in the very darkest skies the naked eye begins to detect some of the individual stars that would otherwise blend together to form the glow. Under excellent conditions, the brightest parts of the Milky Way can even cast diffuse shadows.

Field of View

100° to 50°

(The widest fixed naked-eye field)

THE BIG DIPPER AND THE NORTH STAR

Sight 8

The North Star, also known as Polaris, is the most famous of all night's stars. Its neighbor, the Big Dipper, is the most famous of all star patterns. The Little Dipper is associated with the Big Dipper and so is very well known, too—by name at least, for it is far less conspicuous as a whole than the Big Dipper. Actually, the names of these patterns vary in different countries. In the United Kingdom, the Big Dipper is known as the Plough. In addition, the Big Dipper and the Little Dipper are really only parts of larger, official constellations: Ursa Major the Great Bear, and Ursa Minor the Little Bear. The Big Dipper is the hindquarters (Dipper bowl) and unnaturally long tail (Dipper handle) of the Great Bear. The Little Dipper is the head and body (Dipper bowl) and unnaturally long tail (Dipper handle) of the Little Bear. And where does Polaris fit in to all this? It marks the end of the Little Dipper's handle and end of the Little Bear's tail.

Regardless of what we call them, the fact remains that the North Star and the Dippers are famous not just because of their interesting patterns or their brightness. Neither Polaris nor any of the Dipper stars even belong to the brightest class of stars, the 1st-magnitude ones. As most people know, how Polaris really gains its fame is from marking the direction of due north. And the Big Dipper is famous partly from its virtue of helping us to find Polaris. These traits are especially valuable because, for many observers, Polaris and the Dippers never set. They do not set because they are *north circumpolar*—that is, they make circles around the *north celestial pole*, never dipping below the horizon. The north celestial pole is the still place to which the north end of Earth's rotation axis points. Polaris is located very near the pole. The north celestial pole and Polaris appear as many degrees above the north horizon as is the latitude of an observer in Earth's northern hemisphere (for instance, for an observer at latitude 40° N, Polaris appears 40°—about four widths of your fist held at arm's length—above the northern horizon). Any constellation that is close enough to the celestial pole will travel a circle around it that is never cut off by the horizon. For observers at 40° N or farther north, this is true of the entire Big Dipper (even somewhat farther south on Earth at least part of the Big Dipper stays just above the northern horizon when it passes beneath Polaris and the celestial pole).

Comet Hyakutake passing near the Big Dipper.

Knowing these beautiful and useful relations of the Dippers and the North Star add much to our enjoyment of observing them. But that enjoyment is also based on other, purely visual aspects of them.

Sight of the Big Dipper

We gaze into the northern sky from a fairly dark location well north of the tropics on a clear, moon-free evening. Unless the evening is in autumn, when the Big Dipper is scraping low near the northern horizon, far under Polaris, and may be hidden by trees, buildings, horizon haze, or, worst of all, light pollution, we will gain sight of the Big Dipper right away. Six of its seven stars are of 2nd magnitude (the second brightest class of stars). Only the star marking the juncture of its bowl and handle is decidedly dimmer. No other star pattern except Orion the Hunter and the Southern Cross is both brighter and as noticeable from the strikingness of its form.

We're looking at the Big Dipper now, and even if we stopped here and never identified Polaris or the Little Dipper, we could have a long and wonderful observation.

View each star of the Big Dipper and connect it to its proper name. The names, in order from the handle-end and around the bottom of the bowl to the lip of the bowl are: Alkaid, Mizar, Alioth, Megrez, Phad (or Phecda), Merak, and Dubhe.

If you know a little bit about stars, you may know that most patterns of them are only chance line-of-sight arrangements in the sky: the stars are really at different distances from us and are pursuing entirely different and independent paths through space. Is this the case even with a pattern as dramatic as that of the Big Dipper? Yes and no. The five central stars are actually at a similar distance from us (about 80 to 85 light-years away) and are traveling through space together as the core of a rather loose star cluster. Only the two end stars, Alkaid and Dubhe, are at different distances (102 and 125 light-years, respectively). They are also heading in different directions from the five other stars, guaranteeing that the shape of the Big Dipper will change noticeably in the next few tens of thousands of years.

Dubhe, by the way, is the only one of the Big Dipper stars that is noticeably colored—orange, though you may need binoculars to confirm your naked-eye suspicions of this.

Megrez is the obviously least bright star of the Big Dipper—at magnitude 3.3 it is actually outshined by several other stars in the legs of Ursa Major.

Mizar is the star at the bend in the Big Dipper's handle and is the most observationally interesting. First of all, it is a fine double star in telescopes (see Sight 45). But it is also a naked-eye double star: or, at any rate, Mizar has a companion star that most people with standard vision can see in good sky conditions. The companion, Alcor, is a magnitude 4.0 star that is 11.8' (a bit more than one-third the apparent width of the Moon or the Sun) away from magnitude 2.0 Mizar. Check for Alcor each night to find out how good your sky conditions are and how sharp your eyesight is (if you can't detect it with your naked eye tonight, even the weakest binoculars or finderscope will suffice to reveal it).

As spectacular as the Big Dipper is to see on its own, there's no denying that it takes on even greater fascination and beauty when seen in relation to the North Star, the Little Dipper, and other stars and constellations.

The most famous connection of the Big Dipper is, of course, to the North Star, which can be found using "the Pointers." The Pointers are the two stars on the outside of the bowl—that is, the side of the bowl opposite from the handle. The Pointer in the bottom of the imagined bowl is Merak; the Pointer that marks the lip of the bowl is Dubhe. If you run a line from Merak through Dubhe and extend it just a bit more than one Big Dipper length, it will bring your eye almost right to a star similar in brightness to those of the Big Dipper: Polaris, the North Star.

The Little Dipper and the Wandering Pole

Once you find Polaris, you can look for two fairly prominent stars only about two-thirds as far from Polaris as Polaris is from Dubhe. These two are Kokab, just marginally dimmer than Polaris, and Pherkad, one magnitude dimmer than Polaris. Kokab and Pherkad are the only stars other than Polaris that are conspicuous in the Little Dipper. They mark the side of the Little Dipper bowl that is opposite from Polaris and from the very dim curve of stars that forms the Little Dipper's handle (or Little Bear's tail). Kokab and Pherkad still retain a title that reminds us of their glory days about 2,000 years ago. Together, they are "the Guardians of the Pole." At that point in history, these two stars, and not Polaris, were close to the north celestial pole. They are in fact a visible reminder that our Earth's axis of rotation ever so slowly undergoes **precession**. That is to say, our Earth, like a toy top, wobbles ever so slightly with its axis very slowly describing a vast circle in space—and therefore pointing to one "north star" after another along a huge circle in the heavens. If you want to see the star that stood near true north in the sky more than 4,000 years ago—the early heyday of ancient Egypt—look almost exactly halfway between Mizar and Pherkad. There you will find the Pole Star of the early pharaohs—the modest-looking (magnitude 3.7) star that is known as Thuban. Thuban is part of the mostly dim but large and curving north circumpolar constellation Draco the Dragon—a constellation whose line-of-stars tail separates the two Dippers from each other. Before about 500 B.C., when the north celestial pole began to approach Kokab and Pherkad, Ursa Minor was of less importance and was actually pictured as the wings of Draco. According to legend, it was the early Greek scientist Thales of Miletus who made a separate constellation out of the dragon wings, dubbing them the Little Bear.

The Big Dipper as Sign, Calendar, Clock, and Yardstick

There is so much more you can do with the Big Dipper than just use it as a compass needle, an implement that points to Polaris. The Big Dipper is also a directional sign, a calendar, a clock, and a yardstick.

It is a sign pointing to the brightest star of spring: Arcturus. Just extend the curve of the Big Dipper's handle outward a little more than one Big Dipper length, and you will find zero-magnitude Arcturus (see Sight 15).

The Big Dipper is also a calendar and a clock because its position—above, to the left, below, or to the right—from Polaris tells us the time of year and the time of night.

Finally, the Big Dipper is a yardstick in the sky because it and its parts can be used to estimate angular measure. The entire Big Dipper is about 25½° long. The length of the handle is about 15½°. The distance across the top of its bowl is 10°—about the same as the width of your fist held out at arm's length. The distance between the Pointers is about 5½°. Finally, the gulf between Dubhe and Polaris is 29° and between Alkaid and Arcturus about 30½°.

The Orion Group of Constellations

Sight 9

Many people will tell you that the stars look brighter in winter. Is that because winter nights are clearer than nights at other times of the year? Actually, the time of year with the greatest number of clear nights or the very clearest nights of all depends greatly on where you live. But I don't think either distinction is held by winter anywhere in the United States.

No, the real reason people think the stars look brighter in winter is a simple one: the stars *are* brighter in winter. That is to say, it just happens that by far the most brilliant assemblage of constellations in all the heavens is the one that is high on winter evenings. The brightest constellation of all is at the center of this mighty arrangement: Orion the Hunter. And so I call the brilliant massing of the traditional constellations of winter "the Orion host" or "the Orion constellation group."

Chills of Cold and Beauty

We walk outside in the late evening in December, midevening in January, or early evening in February. Even if the Moon is rather bright or we live in a bright suburb or a fairly sizable city, we will be able to notice that there are glittering stars gathered in the southeastern and southern sky. Let's begin, however, by imagining that we are at a country site on a clear, moonless evening.

There's a chill in the air as we walk out, but when we look up, the shiver that runs down our spine is not from the cold—it's from the startling beauty of bright stars. In one field of view—a large one but still only a small fraction of the entire heavens—hang four of the six, and six of the nine, brightest stars visible from midnorthern latitudes. But it is not just these 1st-magnitude (and brighter) gems that catch our attention. There are also many 2nd-magnitude stars, and all the bright stars are arranged in memorable patterns in this Orion host of constellations. "Host" suggests an army, and Orion, while principally imagined to be a hunter, can also be considered a warrior. And there is indeed a fierce splendor about Orion and the constellations attendant upon him. For one thing, many of his (and some of the other constellations') stars sparkle with a hint of icy blue. That is appropriate for this season of the year and suggests, along with the bite of the cold, sharpness—beauties with an edge more cutting than the frigid winter wind.

Illustrious Members of the Host

In all fairness, we must admit that what seizes an observer's attention first in this giant group of constellations is usually not the group as a whole (unless that is noted briefly): it is the central figure of it, gallant Orion. And first in Orion, we are drawn to his center, his irresistible Belt of three similarly bright stars equally spaced from one another in a short, nearly perfect row. But Orion, and his Belt and Sword regions, and the Great Nebula that gleams in his Sword, are all individual wonders worthy of their own chapters in this book (see Sights 16, 28, and 46).

Did I say that Orion and his Belt steal the show? But what about the star that is by far the brightest in all the heavens (save for our daytime star, the Sun)? That is Sirius (see Sight 14), which marks the head or heart of Orion's faithful hound, Canis Major, the Big Dog. The sixth brightest star visible from midnorthern latitudes is Procyon, the prime light of the otherwise modest pattern of Orion's other canine, Canis Minor, the Little Dog.

Whereas Sirius is lower left from Orion when the Hunter is high, Procyon is almost due left (east) of one of Orion's two brightest stars, the famous Betelgeuse (the other brilliant star is Rigel). To the upper left of the high Orion is one of the sky's other brightest constellations: Gemini the Twins. The heads of the imagined twins are represented by the apparently closest-together pair of really bright stars visible to most of the world's observers: Pollux and Castor, named for the most famous twins of Greek mythology. Pollux slightly outshines Castor. It is one of the least bright of 1st-magnitude stars, whereas Castor is one of the very brightest of 2nd-magnitude stars. The imagined bodies of the Twins extend through a number of fairly bright stars, pointing in the direction of Orion.

Two more constellations complete the inventory of bright ones in the Orion host. Far above Orion is Auriga the Charioteer with its zero-magnitude yellow star Capella. And to the upper right of Orion, supposedly facing the Hunter in fearsome conflict, is Taurus the Bull. Taurus has two stars to mark the tips of his long horns (one of these stars is borrowed by Auriga to help form the desired pentagon of stars that is the main pattern of Auriga). But Taurus draws looks and expressions of admiration mostly because of its two great star clusters and its 1st-magnitude orange star Aldebaran. Aldebaran forms a dramatic V- or arrowhead shape with the large cluster called the Hyades (see Sight 25), even though Aldebaran is really much closer than the cluster. Aldebaran marks the Bull's Eye, and the rest of the V, formed by the brightest Hyades stars, outlines Taurus's handsome face. To the upper right from the Hyades is one of the most lovely sights in all of astronomy: the Pleiades or Seven Sisters star cluster (see Sight 26). The main stars of the Pleiades form a tiny dipper that is much richer than the sprawling Hyades.

Much more information about the individual bright stars and constellations that make up the Orion group can be found in Sights 15 and 17.

Winter's Giant Asterisms

For a long time, observers have noticed that the very brightest stars of the Orion group form an almost symmetrical pattern together. This huge geometry is sometimes called the Winter Hexagon and sometimes the Winter Circle. (There is also a Winter Triangle of Sirius, Betelgeuse, and Procyon, but it fails to be as clearly recognizable as the famous Summer Triangle—see the next chapter.)

The Winter Hexagon is made up of straight lines that run from Sirius to Procyon to Pollux to Capella to Aldebaran to Rigel. The Winter Circle uses the same stars but links them with curved lines. In both of these **asterisms** (star patterns that aren't official constellations), one 1st-magnitude star is left inside the geometric pattern, near the center of them: Betelgeuse. Yet another pattern that uses all these stars—*and* Betelgeuse, near the center—is the Heavenly G. In this case, the space between Rigel and Aldebaran is left open, and a line is drawn from Rigel to Betelgeuse to form the horizontal stroke of the G. Actually, Betelgeuse and Rigel are themselves only horizontal together for midnorthern latitudes when they are rising. So the Heavenly G is usually more or less tipped over.

Whichever of these vast multiconstellation asterisms you prefer, it is best seen when the Moon or light pollution brightens the background sky and wipes out our view of the other less-bright stars in this region. But we star lovers would rather see Orion's host at full blaze in a deep, dark sky.

Sight 10 THE SUMMER TRIANGLE REGION

The great host of winter's bright stars and constellations that surround Orion—the topic of our previous chapter—cannot be surpassed in brightness or number of bright stars. But the Winter Hexagon or Winter Circle of 1st-magnitude stars from those constellations is almost too large, too complex, and too distracted from the neighboring 2nd-magnitude stars and the smaller striking asterisms—especially Orion and Orion's Belt. This last factor—distraction by Orion and Orion's Belt in particular—is what really works against obvious recognizability of what has been called the Winter Triangle. That triangle, almost equilateral, is formed by the brilliant Sirius, Procyon, and Betelgeuse. But neither it nor the Winter Hexagon or the Winter Circle (or the Heavenly G—see the previous chapter) have the simplicity and freedom from numerous other distracting neighboring stars that is possessed by the summer sky's glorious giant asterism: the Summer Triangle.

Ships supposedly get lost in the infamous Bermuda Triangle (the facts don't bear out this pseudoscientific claim). But the Summer Triangle is a place where you can really get lost—in wonder.

The Main Stars and Constellations of the Triangle

In the middle of June's short nights and at the start of July and August evenings, a big and prominent triangle of very bright stars hangs high in the east. Our eyes are first pulled up to the highest and distinctly brightest of the three stars. This is Vega, which falls just short of being the second brightest nighttime star properly visible from midnorthern latitudes. Vega has its hint of blue and its lofty path: this is the bright star that passes almost exactly overhead for observers around 40° N, the most populous latitude on Earth. But there will be time to study Vega and the other two very bright stars of the Summer Triangle by themselves later (in Sight 15). Right now, our vision is drawn well to the lower right (southeastern sky) away from Vega to notice the second star. It is bright, though only about half as bright as Vega, and is quite interestingly flanked closely on either side by a moderately bright star. This second brightest star of the Summer Triangle is Altair. Finally, to the lower left of Vega (when we face east), somewhat closer to it than Altair is to Vega, there is yet another 1st-magnitude star: Deneb. Decidedly less bright than

The Summer Triangle.

Altair and much dimmer than Vega, Deneb nevertheless grabs our attention, especially by virtue of its position in the majestic Cygnus the Swan and its vicinity to some of the strongest Milky Way glow (for information about the Milky Way's Cygnus Star Cloud, see Sight 7).

The three stars of the Summer Triangle each belong to a different constellation. Vega's constellation is Lyra the Lyre (the lyre is an ancient harplike musical instrument). Altair is the leading light of Aquila the Eagle. Deneb is the brightest star in Cygnus the Swan.

Lyra is a compact diamond of modestly bright stars, with Vega mounted on top (the northern end). Lyra really does resemble the tortoise-shell lyre of the greatest musician of Greek mythology, Orpheus, whose soulful playing could make the very stones weep.

But to me Lyra also resembles a little loom, which is appropriate because Vega is the weaving damsel in the greatest and most beloved of all Asian star myths. That legend made its first known written appearance in China over 2,600 years ago and eventually evolved into different versions, especially in Japan and Korea. The legend tells how the damsel (or princess) and the cowherder (or prince) represented by Altair (or Altair and its two flanking stars) fall in love but are so smitten with each other that they neglect their heavenly duties. As a result, the great celestial emperor separates them on either side of a river—the Milky Way, the River of Stars. Sure enough, Vega is on one "shore" of the Milky Way band, and Altair is on the other shore. The lovers are permitted just one night a year together. It is "the seventh night of the seventh moon"—the date is now celebrated in early August—when all the magpies in the world gather to form a fluttering bridge over the Milky Way for the lovers to cross over and be together.

A different time-honored tradition sees Altair as the head or eye of Aquila the Eagle, with the flanking stars like epaulets on the eagle's shoulders. The rest of Aquila's stars do make a rather nice stick-figure form of an eagle.

A more elegant shape is that of Deneb's constellation Cygnus. The line of the Swan's body and its outstretched wings, perpendicular to the body, together make a prominent cross. And this main pattern of Cygnus is indeed also an asterism known as "the Northern Cross" ("northern" to distinguish it from the very famous Southern Cross that is not visible north of the southernmost United States).

Deep-Sky Objects and Little Gemlike Constellations

We'll meet with many of the deep-sky objects of Cygnus and Lyra in upcoming chapters. Anyone looking at the Summer Triangle can't help but be interested in surveying these sights. In Lyra, very near Vega, we find Epsilon Lyrae (see Sight 45), a point of light that can be split into a double star with sharp human eyesight and into four stars by telescopes at about 100× or even less on very steady nights. M57, the Ring Nebula, shimmers like a cosmic smoke ring (see Sight 49). Beta Lyrae is an odd and fascinating naked-eye variable star (see Sight 29). Aquila and Cygnus offer the variable stars Eta Aquilae and Chi Cygni, respectively. But Cygnus is a wonderland of many, many celestial treasures: numerous double stars, including the gold and blue Albireo (see Sight 45); diffuse nebulae such as the North America Nebula (for information on the best of the diffuse nebulae, see Sights 32 and 46); planetary nebulae such as the Blinking Planetary (see Sight 49); and a supernova remnant, the Veil Nebula (see Sight 30).

Lyra, Aquila, and Cygnus are the three constellations that house the three stars of the Summer Triangle, but other wonderful constellations lie within or adjacent to the triangle. Delphinus the Dolphin is like a minute diamond shape of stars with a tail of stars. Sagitta the Arrow is even smaller. Dephinus and Sagitta require fairly dark skies to see well, but just next door to them, within the Summer Triangle, is a small constellation that is more amorphous in pattern: Vulpecula the Little Fox. But within its borders, Vulpecula contains two famous deep-sky objects. One is the renowned Dumbbell Nebula (see Sight 49). The other, located about one-third of the way along the line from Vega to Altair, is an asterism so tiny that it needs to be seen with binoculars or, preferably, a low-power telescope: it is the striking Brocchi's cluster (though it is not a true cluster), also known as the Coat Hanger—for that implement of closets is exactly what the pattern looks like.

Watching the Summer Triangle until Winter

There are so many deep-sky objects in and around the Summer Triangle that you can spend all night touring the region with a telescope. But whether you have an optical aid or not, you will find yourself again and again admiring the mighty Triangle with your naked eyes. Very late on summer nights, or in the evening by September and October, the Summer Triangle passes overhead and begins to creep down the western sky. Whereas Vega was earlier at the top of the pattern, now it is at the bottom, at the same height as the more southerly Altair by the time they are halfway down the western heavens. Deneb lingers above them, with Cygnus the Swan beginning to swirl around to a vertical position: as Cygnus sets, viewers at midnorthern latitudes see it at the top of the Northern Cross pattern, which stands almost straight as it reaches the northwestern horizon. Amazingly, the so-called Summer Triangle remains long after summer. The three-star pattern is last seen, low in the west, in December.

Field of View

50° to 15°

(Moderately Wide Naked-Eye Field)

VENUS AND MERCURY AT GREATEST ELONGATION

Sight 11

Two planets are closer to the Sun than Earth is. This means that we can never see them appear very far away from the Sun in our sky. All the other planets can shine high in the midnight heavens. But these two "inferior planets" (inferior to Earth in their amount of distance from the Sun) can never be seen for more than a few hours after sunset or a few hours before sunrise. The innermost planet, Mercury, can never appear more than 28° from the Sun. The second planet, Venus, can pull out to about 47° from the Sun. These times when Mercury and Venus pull out to reach these maximum angular separations from the Sun are precious times to see them well. When they reach these positions, we say they are at **greatest elongation**.

Observing Venus and Mercury

The Sun has just sunk below the horizon. A gentle spring breeze is blowing. You look up in the west—almost halfway up the sky, though it seems even higher. And what you see there is an intense little point of light, visible before any other planet or star. As the sky behind you darkens, and even where the Sun went down pales from a glare to a vibrant but gentle glow, the point of light gets bigger and more prominent. Even by the time full darkness has fallen, and the light has become a truly impressive beacon, this object seems fairly high, probably still above any buildings or trees in your landscape as long as they are not too close to you. The pure beauty of this sight is great even if you know nothing about it. But its loveliness is appreciated even more if you know a few things. You know that this is the planet Venus and that after many months of its lingering shyly down in trees and bright twilight it has now finally reached its lofty peak—at greatest elongation.

Shift our setting. Now we are out on another spring sunset, seeing another wonder. This time, the thrill is not just partly but mostly in what we know. For unless the sky happens to be exceptionally clear, that speck of light we finally begin glimpsing about 30 to 45 minutes after sunset is rather inconspicuous. Still, it is likely shining brighter than any star that happens to be there low in the west or west-northwest; it is probably the first and maybe only object to

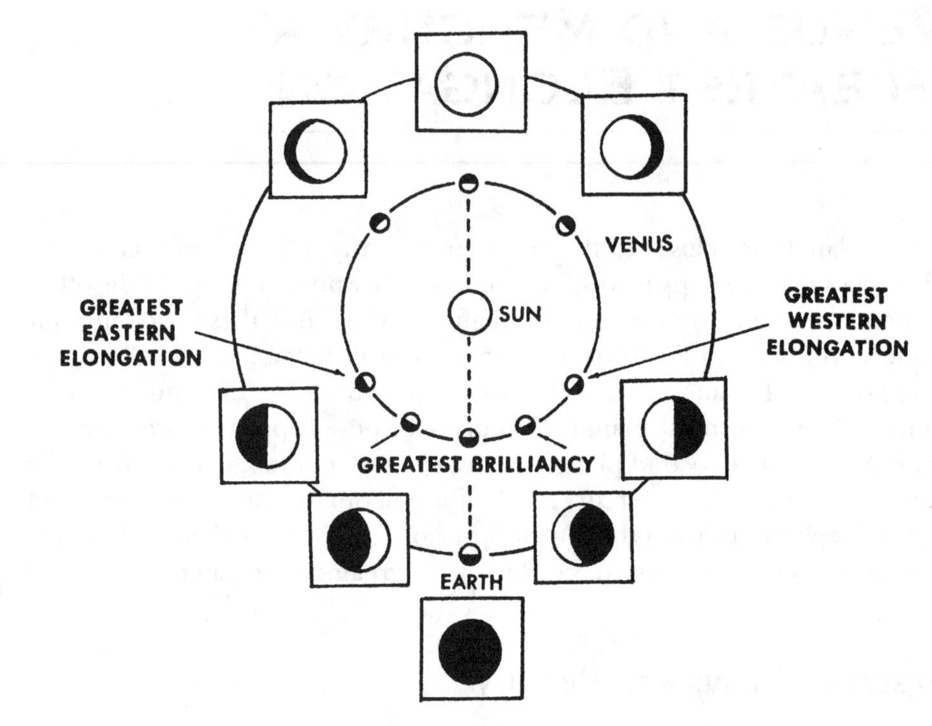

The phases of Venus. The images in the boxes show the appearance of Venus as seen from Earth when Venus is in the various orbital positions shown.

appear in that part of the twilight sky. But what is key is its identity. We are glimpsing a planet that can exceed Sirius (the brightest star) in brilliance, the planet that is the closest to Earth more often than any other—and yet is so elusive that many experienced amateur astronomers have never seen it!

The great astronomer Nicolaus Copernicus, whose work eventually persuaded the human race that Earth and other planets orbit the Sun, is reputed to have never seen Mercury. I know one amateur astronomer who saw, with the help of a rather large telescope, the extremely dim speck of light that is distant Pluto (which was not discovered until the twentieth century). And yet this person had never seen the stealthy world we are talking about. Pluto has never been seen close up by one of our passing spacecraft, but the planet now shining before us has an entire side of it that has never been seen close up and remains a tremendous mystery. What is this planet that is the fastest of all and dodges out into visibility for only a few weeks or even days around greatest elongation a few times a year? It is Mercury.

Venus and Mercury at greatest elongation are not just highly visible gems, they are gems in a vast lovely background of twilight sky and beautifully darkening landscape. They are part of a rapidly changing drama of twilight turning into night.

Actually, around the time of greatest eastern elongation—when Venus is farthest east of the Sun and therefore high in the western sky after sunset—Venus can set three or even four hours after the Sun. Or, a few months later, when Venus has left the evening sky and appears before dawn, it is near greatest western elongation and can rise three or four hours before the Sun. In either case, we can have the luxury of viewing Venus still fairly high in a fully dark sky.

In contrast, Mercury can never set even as much as two hours after (or rise more than two hours before) the Sun. Because astronomical twilight ends when the Sun is 18° below the horizon and the very last hint of its glow disappears from the western horizon, it can never end less than about 1½ hours after sunset. So even when Mercury sets as much as 1¾ hours after the Sun, we can only see it very low after twilight. It doesn't really have a true night part of its stage play. On the other hand, Mercury can change its position and brightness so much in the course of just a few days that this provides additional drama.

Steep versus Shallow Elongations

So far I have been describing what we see of Venus and Mercury at the very best of their greatest elongations. But it's important for us to know that some of the greatest elongations are better for observers than others. There are two major factors that play a role.

The first is the shape of the planet's orbit. In the case of Mercury, its elliptical orbit puts it sometimes much farther from the Sun in space than at other times. Only if we see Mercury at greatest elongation when it is also near *aphelion* (its farthest point from the Sun in space) is the angular separation as much as 28°. If Mercury reaches greatest elongation when it is near *perihelion* (its nearest point to the Sun in space), the angle can be as little as 17°. This factor isn't important at all for Venus because the orbit of Venus is nearly circular.

The second factor that determines how good a greatest elongation is for observers—at least for observers north of the northern hemisphere tropics or south of the southern hemisphere tropics—is the time of year. Or rather, something that varies with the time of year: the steepness or shallowness of the angle of elongation. Around the spring equinox (the first day of spring), the separation between Mercury or Venus and the Sun at dusk can be almost vertical. By contrast, around the autumn equinox the separation is more nearly horizontal—instead of being 47° straight above where the Sun is setting, Venus may be 47° to the slightly upper left of the setting Sun (that is, most of the separation is in degrees to the left of the Sun, not above the Sun). At its greatest elongation around early autumn, Venus might be only 20° or less above the horizon at sunset; therefore, it will be dimmed more by any haze and local light pollution, and more easily hidden by trees or buildings.

The steepness of elongations at dawn displays of Venus and Mercury occurs at opposite times of the year from those at dusk. In other words, the two planets have steep, high elongations at autumn dawns and shallow, low elongations at spring dawns.

How steep or shallow an elongation of Venus or Mercury is depends merely on how far north or south of the Sun they are in the heavens—and how far north or south you are on Earth. It is this factor that can also end up placing these planets at their highest many days or even weeks before or after greatest elongation. Nevertheless, Venus is always at least fairly high and well placed at greatest elongation. And there is a telescopic treat that both these planets always do offer near greatest elongation.

Exciting Telescopic View at Greatest Elongation

This chapter is primarily about the wide naked-eye view of Venus and Mercury seen in the context of earth and sky when these planets are highest in that sky. But we should note that the telescopic view of these planets when they are near greatest elongation is also thrilling. When Mercury and Venus are near greatest elongation, they appear half lit in telescopes: you look in the eyepiece and behold a dazzling (Venus) or a subdued (Mercury) little half-moon shape. Do Mercury and Venus appear *exactly* half lit when at greatest elongation? Almost. In yet another interesting twist, it turns out that the moment when they are at *dichotomy*—that is, they have the appearance of being exactly half lit (a straight line separates their bright from their invisible dark sides)—may be several days before or after greatest elongation. At a particular greatest elongation, on which exact date do you judge the phase of Venus or Mercury to be exactly half lit?

For information on other times when Mercury and Venus are fascinating in telescopes, see Sights 39 and 40. For more on a time when Venus is a spectacular naked-eye sight in another way, see the next chapter.

Venus, Jupiter, and Mars at Brightest

Sight 12

Every month, the Moon goes through its cycle of phases and reaches its brightest at Full Moon. But the brightest planets reach their peak brilliance less frequently. When one of them does, we see—usually for at least a few weeks—their maximum radiance concentrated into a single naked-eye point.

The three brightest planets are Venus, Jupiter, and Mars. Venus hits a peak brilliance of magnitude –4.7 (remember, the lower the magnitude—in this case the higher the negative number—the brighter an object is). Jupiter and Mars can glow as brightly as –2.9, though Jupiter comes very close to this value in only some years and Mars gets anywhere near so bright only a few times every fifteen or seventeen years.

A difference of almost two magnitudes means Venus is about 4 or 5 times brighter than the best that Jupiter or Mars can offer. Venus is thus by far the brightest of the bright. On the other hand, Venus can never be seen high in a midnight sky—as the other two can when they reach maximum brightness. Furthermore, each of these worlds is different and has its own special characteristics. For example, although Mars can go many years before it passes close enough to Earth to rival Jupiter in brilliance, it also has a more distinctive and impressive color than Jupiter.

The Sight of Brightest Venus

Let's start with the peerless lamp of the twilight and night, the planet Venus. So prominent is this lanternlike planet that in the lore of many cultures it is part of a trio of the qualitatively most important celestial objects along with the Sun and the Moon. When Venus is visible after sunset, it is often and poetically called "the Evening Star"; when it is visible before sunrise, it is often known as "the Morning Star."

Our scene is similar to the one we painted in the previous chapter, where we considered Venus at greatest elongation from the Sun. Venus becomes its brightest about five weeks after it reaches its greatest evening elongation and about five weeks before it reaches its greatest morning elongation. These are the times when the combination of Venus's size and phase—both discernible in telescopes—produces the most light. The side of Venus facing us is then about one-fourth illuminated (a moderately thick crescent) and the disk diameter (measured from one Venus crescent point to the other) measures about 43".

But let's return from the telescope to our naked-eye view at maximum brightness. We see the planet typically lower than it was at greatest elongation. It is, however, noticeably more luminous. How bright is Venus within a week or two of its "greatest brilliancy"? No matter how many decades I have watched the heavens and followed Venus and marveled at its brightness, its radiance at greatest brilliancy has never ceased to strike me with awe. It is the extra radiance that just beggars belief. Of course, this is much more the case when you see Venus on a moonless night at a dark country location. Under such conditions, shadows cast by Venus can be seen on white surfaces. When I was a kid, I was once startled to see a long pathway of light running across a field covered with an icy-surfaced layer of snow. The source of the prominent light was Venus shining through a break in the trees.

Now, city lights are so wastefully bright and misdirected that urban observers have to settle for the fact that Venus at brightest actually competes with many (relatively distant) lights in the cityscape. Like all outer space objects, however, Venus is so distant that it seems to move with the observer relative to nearby objects, including artificial lights. Therefore, it stands out—or, we should say, "moves out"—from among the streetlights and other light sources of the city.

When Venus is brighter than usual, it is often reported as a UFO. If the observer is moving, he or she thinks the object, possibly an alien spaceship, is pursuing him or her. If the observer is still, then Venus is thought to be hovering and, then, over the course of many minutes, slowly sinking from view beyond that ridge or forest (of course, in reality, the sinking is not the plan of an extraterrestrial pilot, it is just the rotation of the Earth that makes all celestial objects appear to rise and set). During World War II, battleships fired on Venus when it came out from behind clouds, resembling the bright light on an enemy aircraft.

When Venus is brighter than usual, it is sometimes noticed in broad daylight by the general public. This happened once when a crowd in Paris was distracted from the haughty Napoleon by a strange point of light in the blue sky—Venus.

A few weeks after it attains greatest brightness in the evening sky, Venus becomes a much thinner crescent in telescopes and loses a lot of brightness. It starts setting so much sooner after the Sun that it becomes lost in the solar afterglow, and then it re-emerges in the following weeks to rise sooner and sooner before the Sun, until it reaches greatest brightness again, only about twelve weeks after the evening brightest. Unfortunately, then we have to wait about fifteen months before Venus is again at maximum light, again in the evening sky.

What is the secret of Venus's tremendous brightness? The high reflectivity of the clouds that always totally enshroud it. The clouds of Jupiter are almost as reflective, but Venus has the great advantage of being closer to the Sun. The clouds of Venus are therefore lit with much brighter sunlight.

The Sights of Brightest Jupiter and Mars

As we noted earlier, Jupiter and Mars may be far less bright than Venus, but they do have the advantage of being visible at their highest in the middle of the night. The reason for this difference is simple. Venus is one of the **inferior planets**—inferior in the sense that its distance from the Sun is less than Earth's. So Venus can never be seen at a very great angle away from the Sun by people on Earth. Mars and Jupiter, however, are **superior planets**—superior in distance from the Sun compared to Earth. We can sometimes look straight out away from the Sun and see Mars or Jupiter in that direction. When a superior planet reaches a location directly opposite the Sun in our heavens, that position is called **opposition**. The planet at opposition rises at sunset and is visible all night until it sets at sunrise. Opposition is also when Mars or Jupiter—or any of the superior planets—is approximately closest to Earth and looks biggest in telescopes. Opposition occurs when a superior planet is brightest, not only because its disk looks biggest in telescopes then but also (to a lesser extent) because the side of the planet facing us is also facing the Sun and is therefore fully lit.

Jupiter appears yellow-white, a color very similar to that of Venus. Jupiter, like all sizable planets, shines with a steady light. It is also the most steady and dependable of the planets in its performance. By this I mean first of all that Jupiter, though distant, is so big in true size that it always looks bright. Furthermore, Jupiter happens to take about twelve Earth-years to orbit the Sun once. This means that it spends about one Earth-year in each of the twelve classic constellations of the zodiac. You don't have to worry about Jupiter flying onward to other constellations in a matter of weeks (as Venus and Mars can) or taking years to creep out of a single constellation (as Saturn and even more distant planets do). In the course of the twelve months that Earth requires to orbit the Sun once, Jupiter has advanced almost one-twelfth of the way around its own orbit. So it takes an additional month for Earth to draw even with Jupiter again. In other words, each year the opposition of Jupiter comes about a month later.

Mars is the third and last planet that can get much brighter than even the brightest star. Whereas Earth takes only thirteen months to come back around its orbit and pass between Jupiter and the Sun (placing Jupiter at opposition), Earth takes much longer to catch back up to Mars. After all, Mars is not that much farther out from the Sun than Earth is, so its orbit is not vastly larger than ours and its orbital speed not greatly slower than ours. Oppositions of Mars come only about once every twenty-seven months—so roughly once every other year.

But there is a further twist about oppositions of Mars and how bright the planet gets. Jupiter has a nearly circular orbit, so whether we catch up to it near its *perihelion* (its point closest to the Sun in space) or *aphelion* (its point farthest from the Sun in space) makes only a mild difference. Jupiter appears somewhat brighter at a perihelic opposition than it does at an aphelic one—but not spectacularly brighter. In striking contrast, Mars has an orbit that is decidedly elliptical. Mars can come almost twice as close to Earth at a perihelic opposition than it does at an aphelic opposition. Mars can appear two magnitudes brighter and, in telescopes, more than a third wider at perihelic versus aphelic oppositions. The progression of oppositions of Mars is a dramatic one. After a stunning perihelic opposition, the oppositions—coming about once every two years—get steadily more distant and dimmer for either about seven or nine years. Then the process reverses and each opposition gets brighter until—either fifteen or seventeen years after the previous perihelic opposition—another perihelic opposition occurs. Thus, in 2003 Mars had a perihelic opposition (in fact, Mars was then slightly closer to Earth than it had been in over 59,000 years!). It shined at a magnitude of –2.9. In 2012, it will be at aphelic opposition and glow no brighter than –1.0.

Of course, even magnitude –1.0 is impressive—especially when it is combined with the color of Mars—the golden orange of a campfire or a pumpkin. No other planet has so distinctive a hue. The brightest golden orange star is Betelgeuse, which at an average magnitude of +0.5 is much dimmer than Mars is even at an aphelic opposition. Furthermore, while Betelgeuse twinkles, Mars shines with a steady, fixed light. And whereas subtly colored Jupiter stares at us, fire-colored Mars positively glares at us. When Mars has one of its closer oppositions, it is not just an impressive sight; it is truly formidable.

BRIGHT COMET WITH LONG TAIL

Sight 13

Of all the wonders in the heavens, there are two that have inspired fear and awe more often than any others: eclipses and comets.

What is the competition for these two? Meteor storms? Most people who don't follow astronomy never get to see a meteor storm or a very strong meteor shower, for they are rare events and hard to predict accurately. Great auroral displays? Throughout history, not many people have seen such displays unless they lived at a fairly high latitude (though nowadays those of us farther south can often be forewarned over the Internet and catch a truly splendid aurora at least every few years).

Eclipses have had a much greater impact on people's consciousness. A lunar eclipse typically lasts longer than a meteor storm or individual auroral displays and is easily noticed. Total eclipses of the Sun are brief but so spectacular they are impossible to miss if you are outside or inside near a window. Of course, they are rare for any given location on Earth—but their fame spreads far and wide so that people even travel to witness one, for they can be predicted far in advance.

It's no wonder then that shocking midday darkness and "the Eye of God" (a total solar eclipse) or a dramatic dimming and reddening of the Moon (a total lunar eclipse) have aroused fear and wonder from earliest times. But why is it that comets have grabbed as much attention and perhaps been deemed even more ominous than eclipses?

Why Comets Were Feared

There is a central reason that comets were feared by our ancestors: of all the heavens' major kinds of sights, comets seemed to be the only ones that were spectacularly unpredictable and variable.

The stars always maintained their patterns. The Moon and the planets roamed from one constellation to another but always the same constellations (those of the zodiac) and always in the same way with completely predictable changes in appearance (the Moon going from crescent to full and back; planets brightening somewhat over the course of months, then slowly fading). Even meteors, though impossible to forecast individually, happened often enough for their behavior to become familiar. But not comets.

Comet Hyakutake.

A bright comet could appear at any time in any place in the heavens and remain visible for weeks while engaging in strange motions. It could keep changing its appearance—drastically—and look different from any other celestial object, even other comets.

Comets are the most prodigious and mysterious objects in the solar system. And whereas eclipses, meteor showers, aurora, and other celestial spectacles tend to last mere minutes or hours, a great comet may remain visible to the naked eye for several months and awesome for several weeks or longer. A total eclipse of the Sun packs its staggering power into so short a time that it leaves one virtually breathless. But a great comet, while somewhat less intense, can spread its beauty over many nights and during this period transform through many variations of loveliness.

The Parts of a Comet

Most bright comets have at least two major parts: the fuzzy point or patch of light called the *head* and an appendage of softer radiance called the *tail*. There are two most important kinds of comet tails. One is a usually straight streamer of light (bluish on photographs but only rarely bluish visually) called the *gas tail*. This is composed of gas that shines on its own when stimulated into fluorescence by atomic particles in the *solar wind*, which pushes the gas rapidly away from the head in the direction opposite the Sun. The second major kind of comet tail is a broader, curving fan of radiance (yellow-white or even slightly reddish on photographs but usually only slightly colored even in binoculars and telescopes). It is called the *dust tail*. It is composed of dust that shines by reflecting sunlight. Though the dust particles are tiny, they are far bigger than molecules of gas and get pushed more slowly away from the Sun—by the pressure of solar radiation.

Often, we see a comet and its tail not far out of the plane of the comet's orbit, so that even a broad and curved dust tail looks straight and narrow (the dust is mostly confined to near the plane of the comet's orbit, not getting very far above or below that plane). In such cases, gas and dust tails may be seen superimposed and blend together from our point of view. On the other hand, if we see the comet and its tails more broadside, they can look something like a giant checkmark, with the curving fan of the dust tail well separated from the straight gas tail.

Glimpses of the Great Comets

Every bright comet is different, so it is really only fair for me to describe not a generic one but specific ones I have witnessed. I have written at much greater length about them elsewhere (especially in my book *Comet of the Century*). Here, I will offer only a synopsis of some of their most outstanding visual aspects. Bear in mind that these were only the comets I felt to be the most spectacular and interesting of all. If by bright comet we mean one fairly easily visible to the naked eye from a dark site, then an example of such a comet tends to be visible at least every few years. What follows are only the best of the best.

Comet Bennett (1970)

I was young and had never seen a truly great comet before this. My home skies were also still pristinely dark. But even taking into account these advantages, it's possible that Comet Bennett was still simply the most beautiful of all the twentieth-century comets.

My first view of Bennett was of it rising above a tree line in bright morning twilight with its head like a 1st-magnitude or brighter star and an intense spike of tail plainly visible to my naked eye, even though the dawn glow had overwhelmed almost everything else. But in the weeks that followed, Comet Bennett kept rising longer and longer before sunup and getting higher and higher in a fully dark sky. Most of its 10° or longer tail was so bright that it could be glimpsed right out any window. In this very important respect, it was, like 1976's Comet West, better than the great comets of the 1990s. Outdoors, Bennett truly looked like a white angel or a sword hung beside, and outshining, the Milky Way's bright Cygnus Star Cloud—or maybe it was an emblematic giant swan's feather of light next to Cygnus the Swan. Even in my fairly small telescope, the comet's tail seemed to consist of silky strands dusted through with sparkling stars (Bennett possessed both a dust tail and a very unusual curved or bent gas tail). Higher in the sky and deeper into the night the comet got, fading as it passed north toward the bright zigzag of Cassiopeia.

Comet West (1976)

Cloudy skies kept me from seeing Comet West's head as bright as Jupiter in bright evening and then morning twilight (the expert observer John Bortle even glimpsed Comet West's head with his naked eye a few minutes before sunset one day). When my skies finally cleared, I stepped out a few hours before sunrise and whimsically told myself that perhaps the comet's tail was so long and prominent I would already see it sticking up from the tree line. Then I stopped dead in my tracks. For there *was* a veritable beam of light extending up from the forest. And it was indeed the tail of Comet West. In a while, as dawn began to brighten, I finally saw, through a break in the trees, the comet's head and the brightest part of its tail. I saw most of the wondrous, grand details of these with a finderscope and a regular telescope, and I will mention a few in Sight 24. But the naked eye showed me much. It showed me the fuzzy spot of the head containing an intense starlike point of light. What's more, the structured first section of the tail shined a vivid gold even without optical aid. The rest of Comet West's tail became more readily visible and understandable to me in the week that followed, as the comet climbed higher into night. I saw West broadside enough for its nearly 20°-long straight gas tail to appear separately to the right of its more than 20°-long and awesomely broad fan of dust tail. The unaided eye couldn't see as much detail in that vast plume of sunlit dust as photographs show but could discern dimly several branches and features of it.

Comet IRAS-Araki-Alcock (1983) and Comet Halley (1985–1986)

While neither the briefly near-passing Comet IRAS-Araki-Alcock (IAA; the closest comet in more than 200 years) nor the classic, famed Halley ever displayed tails that were both long and bright, superb sky conditions on key nights enabled me to glimpse thrillingly great lengths of dim tail on them. Comet IAA's head appeared amazingly big, albeit of rather low surface brightness, on the nights it was nearest. This was especially true on the night of close approach when it floated near M44, the big Beehive Star Cluster, which the comet's head looked like a much bigger version. What about Halley's Comet? The apparent surface brightness of its tail increased as our angle of view on Halley's dust tail changed rapidly to fewer broadsides in late April/early May 1986. As a result, I was able to glimpse more and more of its length. I still hope to confirm someday that I really did trace out that tail's length to 50° (if stood on end, the tail would have stretched more than halfway from horizon to overhead).

Comet Hyakutake (1996)

The comet authority John Bortle announced he believed we were in for a display of historic proportions as Comet Hyakutake headed toward us—for a while almost directly toward us. Its head kept growing and brightening and then started marching rapidly north. Suddenly, one night it came out of the clouds and hung high over my house in the middle of the night with a bright few degrees unfurled like a banner and much longer section of dimmer gas tail pointing slightly downward (an orientation very hard to get from a comet). As Hyakutake came nearest Earth, its head became a remarkably big ball of glow, reaching zero magnitude on a few nights when it was visible all night long and passed virtually overhead for much of the world's population. Its tail on those nights was not bright for anywhere near as great a length as those of Bennett and West, for it was then largely a gas tail (which doesn't glow as bright as a strong dust tail at the wavelengths to which the eye is most sensitive). But Hyakutake's head and a little of its tail were bright enough to show apparent blue to most eyes at dark country sites. Photos show it looking

Comet Hyakutake and its streamers.

like Luke Skywalker's blue lightsaber as it passed the Big Dipper, growing first as long as and then much longer than that large star pattern.

On the night the head passed near the North Star, the tail was almost like a vast clock hand swinging around from the pivot of the star. That night, under excellent observing conditions, I could trace the faintest extension of Hyakutake's tail out to over 70° long (a few skilled observers in ideal climates saw a substantially longer tail than even I did that night!). As Hyakutake rapidly receded from Earth and approached the Sun, its apparent brightness for those of us on Earth decreased, and its tail dwindled—but that tail did spout a strengthening dust component that was pretty as the comet passed from view into the solar glare.

Comet Hale-Bopp (1997)

This famous comet, seen by millions of people, was notable for remaining bright enough to be seen in most amateur telescopes for about two full years—and brighter than zero magnitude for about two full months. It first became a really bright naked-eye comet in the February predawn sky, where it became a fourth component added to the Summer Triangle (see Sight 10). It quickly grew brighter than first Deneb, then Altair, then finally Vega. Its tail was not extremely long, but the first few degrees of it were bright enough to see until 2nd-magnitude stars faded out in dawn. As Hale-Bopp passed north and into the evening sky, it reached its peak magnitude of about –0.8 and

Comet Hale-Bopp.

displayed truly spectacular hoods of light in its head for viewers with optical aid (see Sight 24). In early April, the comet's head and the first few degrees of its tail were bright enough for me to see right over the headlights of heavy auto traffic in the opposite lane as I drove. Hale-Bopp was then setting after the end of evening twilight, and its widely separated dust and gas tails were plainly visible for more than a few degrees and faintly for up to 20° or more under superb conditions. Hale-Bopp was still a bright naked-eye comet when it got lost in the solar glare after sunsets in May.

Getting the Word

To see a bright comet, you first must know that it exists. Unless the comet is extremely bright and very conveniently placed in the early evening sky, there will probably be little about it in the general news media (newspapers, television news, and Internet news services). Before it reaches the climax of its display, would you see an article about it in your favorite astronomy magazine? Probably. There is usually quite a bit of advance warning that a new comet is going to get bright.

But the really devoted comet observer has to be extra vigilant. Occasionally, a comet may be discovered when it comes out from our line of sight with the Sun and is already brilliant. Or an already known comet can sometimes flare up unexpectedly, surprising everyone by getting many times brighter almost overnight. Many years ago, the only way you could keep posted from day to day on the latest comet discoveries was by having a rather expensive subscription to International Astronomical Union circulars. The circulars usually announce professional observations and measurements of celestial objects too faint for the amateur to hope to see—if the object even emitted any significant light at visual rather than radio or infrared wavelengths.

Now, of course, there are a number of comet Web sites that you can check for discoveries and other timely information as often as you like. Or you can subscribe to the free Astroalert service. See the sources section for details.

Once you know a bright comet is visible, and where and when you can observe it, you might not need any special circumstances to detect the comet's head and brightest tail. But comets are usually at their brightest when relatively near the Sun in space and therefore in our sky. Bear in mind that if a comet is low in bright twilight, you will see it much better if the humidity is low and the sky is relatively free of haze. Furthermore, if a bright comet has a long tail that extends far up into the night sky, well out of twilight, you will want to observe from a rural site with a dark sky so you can see the farthest and faintest extensions of the tail.

SIRIUS, THE BRIGHTEST STAR

Sight 14

The first section of this chapter is devoted to a part explanation/part hymn of praise about what any astronomer would regard as the most essential or integral kind of macroscopic object in the greater universe: stars.

The Beauty of a Star

Sadly, many people in modern society have lost touch with the stars in the sky. Certainly light pollution has played a major role in this by degrading our view of stars in and around cities. Whatever the causes, when most people hear the word *star* now they are likely to think first about entertainment celebrities. Yet, the fact that society calls its most figuratively illustrious and admired of personalities *stars* suggests that the original stars—the enduring ones in the sky—were originally objects of our highest esteem.

They should be. After all, what is more truly beautiful and admirable than a star? We are creatures who physically need light, almost as fundamentally as do plants, which are powered by it. We are also thinking and feeling beings who are absolutely cheered by light. So what is more lovely than a speck in great darkness that seems entirely devoted to giving forth light—in all directions and nearly forever? The life of most stars is measured in billions or tens of billions of years. The stars have an ancient majesty older and greater than anything on Earth. And yet a star is also as immediate a thing as there is: we see it tremble—almost as if it is alive—as it shines, right now. Stars thus partake of and present us with both eternity (or near-eternity) and the moment.

Now you may be thinking that if we are light lovers, the Sun coming up at night's end must be the most optimistic and powerfully beautiful of sights. But the Sun itself is a star, merely one tremendously closer to us than all the others—for it is *our* star. The Sun is admittedly the mightiest of stars in its appearance to us, who are not light-years but only 8 light-minutes away from it.

Even so, the nighttime stars, however much stupendously dimmer than the Sun, have the collective advantage of being multitudinous. And even individually, each nighttime star has the advantage of being dim enough to allow us to stare directly at it and its flickeringly changing beauty—a perfect opportunity for contemplation. Each star is also dim enough to allow us to appreciate its setting in what is perfectly and beautifully its opposite: the darkness of the universe around it.

I wrote in Sight 1 about what is probably the most fundamentally beautiful sight in astronomy, along with the Sun and the Moon (but the Sun and the Moon are so much a part of our lives they seem objects of more than astronomy). The sight in question was a naturally dark and clear sky brimming with stars. But the closest rival in fundamental beauty for all the sky's stars seen at once must be this: a single bright star, preferably the brightest of night's stars, seen in naked and pure beauty in a glorious setting of sky and landscape.

The brightest of night's stars, not just by a little bit, but by one or even two qualitative levels of brightness, is Sirius.

Three Sights of Sirius

I've stared at Sirius countless times in my life. Its beauty for me never fails. When I think of Sirius, though, I think of three both wonderful and emblematic sights of it I've had.

The first sight came on the very first night that I ever looked at the stars with knowledge. I believe I had just turned six years old. I had read a book that had a section devoted to identifying certain stars—the brightest stars of winter.

The book talked about the brightest constellation, Orion the Hunter, and its brightest stars, blue-white Rigel and orange-gold Betelgeuse. And it talked about the even brighter star in one of Orion's hounds, the constellation Canis Major, the Big Dog. That star, it said, was the brightest of all, and it was called Sirius. The statements in the book captured my young imagination.

That night, a winter night, I lay in bed staring out through the slightly open Venetian blinds of my bedroom window. Suddenly, I caught sight through the blinds and through breaks in the bare branches of the forest outside of a beautiful bright blue-white spark of a star. This, I excitedly thought, must be Sirius. I'm not sure how long I laid there gazing at that point of light in rapt wonder and pondering. I think it may have been at least 20 or 30 minutes, because I remember there being a substantial change in the position of the star in relation to the tree. But all at once something startled me: well to the left of the star, another light source had come into view. It had a bluish tinge and it was *very* much brighter than the first star. Immediately, I realized the first star had been Rigel. The second—bright beyond bright, a wonder whole orders of greatness beyond that of Rigel—was Sirius.

Now I leap ahead quite a few years to a New Year's Eve. My brother and I were at our family home and wondered about the best way to bring in the new year. Passionate about astronomy as I was, I had an astronomical idea. We went outside. Perhaps we brought flashlights; I'm not sure. But if we did bring them, we tried not to use them. We made our way slowly along a winding path that had been cut as a firebreak in our woods. We walked a few hundred yards, no more, as midnight and the transition to the new year approached—but it could have been a thousand light-years or an even greater distance immeasurable in physical units. We checked a watch we had. The time had come. We walked off the heavily tree-surrounded trail into a very small clearing I knew about. A larger section of the starry sky opened over us in majesty. And there, at its highest in the south, its midnight **culmination**, was the seeming ruler of that majesty: Sirius. In our era of history, this star of stars is at its peak height at the very minute the new year begins.

There is one last remembered sight of Sirius I want to relate. I've actually had this experience with Sirius countless times, and you can, too. But this particular night I was young, and my heart was already uplifted by some thoughts, hopes, and dreams, and I remember distinctly where I went to stand and what I witnessed. Sirius was still fairly low at that hour, so I had to go to the northern edge of our property to get it over the tree line. When I did, I was smitten by the pure and simple beauty of its light, but there was something more about it: it was twinkling unusually strongly and from its heart was shooting out, in unpredictable but therefore all the more mysterious progression, here and there, every few seconds, one ray or pulse after another of variously every color of the rainbow.

Actually, any star can show these marvelous flickers of different colors. They occur when *scintillation*—the technical term for twinkling—is especially strong because the atmosphere is unusually turbulent. The effect is greater when a star is low in the sky and therefore having to shine through a longer pathway of the unsteady atmosphere. Why then do casual skywatchers almost never notice the color changes with any star other than Sirius? Because only Sirius is bright enough for the color changes to be prominent to the naked eye.

The Dog Star and Its Companion

Why is Sirius the brightest star in the nighttime sky? Partly because it is close, as stars go—only 8.6 light-years away. It is the second closest star visible to the naked eye (people who live far enough south can see the even closer star Alpha Centauri). However, Sirius is also the star of greatest true brightness

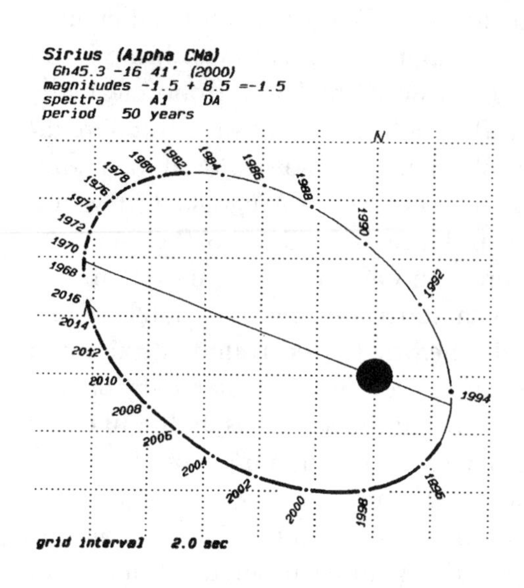

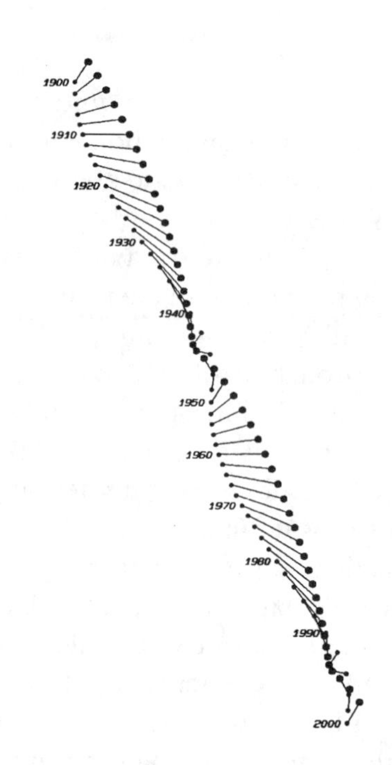

The orbit and braided paths of Sirius A and Sirius B.

that is less than 25 light-years distant. It shines with about 22 times the luminosity of our Sun and is much hotter. But what is most interesting of all about Sirius as an object in space? The fact that it has a **white dwarf** companion.

Whereas Sirius is estimated to be 71 percent wider than the Sun (and therefore almost 1½ million miles across), its white dwarf companion is probably a little smaller than Earth (and therefore less than 8,000 miles across). The white dwarf, Sirius B, is about 10,000 times dimmer than Sirius A. Yet we know that the mass of Sirius A is almost exactly twice that of the Sun and the mass of Sirius B is almost exactly the same as the Sun's. Sirius B is thus incredibly dense—on Earth, a thimbleful of its material would weigh about a ton. This little star packs as much material as the Sun has into a volume only about one ten-thousandth that of the Sun.

The average distance in space between Sirius A and Sirius B is roughly the gulf between the Sun and Uranus in our own solar system. We see the gap between Sirius A and Sirius B increase and decrease during the course of Sirius B's fifty-year orbit. But the apparent separation is always small enough to make detecting Sirius B at least somewhat difficult, even with large amateur telescopes. This was particularly true in the 1990s when Sirius B was situated closer than usual to Sirius A. By 2006, the separation between the two stars was much larger, enough to make it a possible target for an excellent 8-inch or 10-inch telescope on a night of steady, superb seeing. The gap between the two stars will increase even more in the decade ahead.

Sirius is the brightest of many bright stars that make up the constellation Canis Major. It is usually depicted as marking the head or heart of this hound of Orion the Hunter. And this canine connection of Sirius has earned it the title "the Dog Star." As a matter of fact, the term *dog days*, which we still commonly use to describe the hot, humid days of summer, is derived from a fallacious belief that ancient cultures had about Sirius. In classical Greek and Roman times, skywatchers noticed that Sirius kept setting sooner and sooner after the Sun in spring until it was lost in the Sun's afterglow. But they also knew that the *heliacal rising* of Sirius—its reappearance in the sky just before sunrise—occurred in late summer. They reasoned (correctly) that in early to midsummer Sirius must be accompanying the Sun in the daytime sky (though of course invisible in the bright blue daytime sky). They further reasoned (incorrectly) that Sirius was so bright that it must also be warm enough to create the year's hottest weather when added to the Sun's heat in July and August. In reality, even though Sirius has a much higher surface temperature than our Sun and is larger, emitting more total light and heat, it is much too far away to have any significant heating effect on Earth.

By the way, Sirius is one of the few celestial objects other than the Sun and the Moon that has helped form the basis for a calendar. The ancient Egyptians marked the beginning of their new year by the heliacal rising of Sirius. This rising was important because at one stage in Egyptian history it signaled the coming of the annual flood of the Nile—a life-giving event because it rendered fertile what would otherwise be dry desert soil near the river.

One last fact about Sirius is very much worth pondering whenever we observe it. Just as summer's brightest star Vega is located near the point in the heavens toward which our solar system is heading, winter's supreme star Sirius is near the point from which our solar system is departing. Thus, when you look at Sirius, you are in a sense looking out the back window of the solar system, gazing back to where we came from.

Other Bright Stars

Sight 15

In the previous chapter, we discussed Sirius, the supremely bright star of the night sky. No other star is as visually impressive to the naked eye. Nevertheless, each of the other 1st-magnitude stars has considerable brilliance of its own, combined with other distinctive or even unique attributes that make it interesting in its own right. And not just interesting to think about. The different colors and positions of these stars makes each of them very much worthy of actually observing. And although looking at them close up, in a narrow view through binoculars or a telescope, is exciting, be sure to see them as part of a wider view, within the context of their entire constellation and an even larger panorama of stars and landscape. (For maps of some of these panoramas, see the seasonal maps in Sight 1.)

The Brightest Stars of Winter

We already saw, in Sight 9, that winter's Orion group of constellations is home to many of the brightest stars. Seven of the fifteen 1st-magnitude stars visible from around 40° N latitude are found in this one season's group of constellations.

After Sirius, winter's brightest star is Capella, the chief luminary of Auriga the Charioteer. This zero-magnitude object shines with a slightly yellowish tint. The name *Capella* means "she-goat," because the star is imagined to represent the mother goat being held by the Charioteer. Also in his arms are three baby goats—"the Kids," three stars of moderate naked-eye brightness that form a little triangle near Capella. The point of light we see as Capella is actually caused by two yellow stars that are much bigger and more luminous than our own Sun, though of similar spectral type to the Sun. Study of Capella's spectrum reveals its *duplicity* (doubleness) because the two stars are only about as far apart in space as are the Sun and Venus; as such, they are too close together to split visually from 42 light-years away on Earth.

Ever so slightly dimmer than Capella is Orion's brightest star, Rigel. It is the classic example of a **blue giant**, a very hot blue-white star of great size and luminosity. Rigel was once thought to be decidedly more luminous than any of the other mighty blue giants that populate the Orion region. We now think it is a little closer to us—about 800 light-years away—than previously estimated. So perhaps it is only one of the few most luminous stars in the winter sky. Even so, that means its true light output is tens of thousands of times greater than that of our Sun.

Next in apparent brightness among winter's blazing jewels is Procyon, the only prominent star in Canis Minor, the Little Dog. Procyon might be ever so slightly yellow in binoculars or a telescope. Although it is the sixth brightest star of any season visible at midnorthern latitudes, it is often overlooked because of its relatively uninteresting little constellation and the fact that it is so greatly outshined by Sirius, the beacon in the Big Dog. Procyon comes in second to Sirius in closeness for Northerners (it's very close to us, but 11.4 light-years to Sirius's 8.6) and has less true brightness.

Orion's most famous star is not Rigel, but Betelgeuse. Betelgeuse is renowned for its strange name (most astronomers pronounce it *BET-el-joos,* not *beetle-juice*). But its scientific fame comes from it being the classic **red giant**. It is the red giant of greatest apparent brightness, and although research of recent years suggests that summer's Antares may be larger, it is a huge star, comparable in size to the entire orbit of Mars. For observers, it is the Mars-like color of Betelgeuse—a fiery golden-orange—that is most interesting, especially when it is contrasted with blue-white Rigel. But observers should also always check the brightness of Betelgeuse. It is the brightest of all markedly variable stars. Its average radiance is about magnitude +0.5, a little fainter than Procyon. But it can sometimes appear somewhat dimmer or considerably brighter—in fact, in one or two winters a lifetime, Betelgeuse has actually rivaled or surpassed the apparent brightness of Rigel and Capella. This star is also a prime candidate to blow up as a supernova. This is not likely to happen for thousands of years or much longer, but it's not inconceivable that it could blow up in our lifetime. If it did, it would outshine the Full Moon for a few months, and its harder radiation might pose a moderate danger to living things on Earth.

Aldebaran is almost always dimmer than Betelgeuse, and its lighter orange tint is a bit harder to notice with the naked eye. But this is the only 1st-magnitude star that appears right at the edge of a large, bright star cluster. Note the word *appears.* Aldebaran looks like it is part of the huge V-shaped Hyades star cluster, but it is really much closer than the cluster members (see Sight 25). The cluster forms the face of Taurus the Bull, and Aldebaran is the bright eye of the Bull. A line drawn to the upper right from the Belt of Orion carries the eye in the direction away from Sirius and almost right to Aldebaran.

Pollux is the brighter of the two bright stars that mark the head of the twins in the constellation Gemini the Twins. The bright star is just 4½° from Castor, about a half-magnitude dimmer than Pollux. Pollux is slightly orange to the naked eye, more definitely so in binoculars. Castor is just a bit too dim to be a 1st-magnitude star, but a good small telescope can split it into two close-together white sparks—and there are actually at least six suns in its system!

The Brightest Stars of Spring

Spring's brightest luminary is far less brilliant than Sirius but slightly brighter than winter's Capella and Rigel and summer's Vega. It is Arcturus, whose color is usually listed as orange but is actually very unusual and distinctive (a friend of mine once called its hue "champagne shot with roses"). Arcturus can be located by extending the handle of the Big Dipper outward about one Big Dipper length. The star is the only very bright one in the constellation Bootes the

Herdsman. It's been said that this herdsman protects the flock of all the other constellations from Ursa Major, the Great Bear, the giant constellation of which the Big Dipper is just a part. Old traditions consider Arcturus itself the herdsman and, indeed, its name literally means "bear guard." Arcturus is related to the word *Arctic,* which translates as "region of the bear," possibly meaning that the Great Bear constellation is very high and prominent in the Arctic.

There is, however, another high place that Arcturus is associated with: it comes from the region far above the equatorial plane of our lens-shaped galaxy. All the other 1st-magnitude stars travel more or less in the galaxy's equatorial plane, but Arcturus travels on a more inclined orbit. Most other currently brilliant stars would have been pretty bright a half-million years ago, but Arcturus was then so much farther from us that it was only dimly visible to the naked eye. We happen to live in the time when it is diving almost its nearest to us down south through the equatorial plane (the current distance of Arcturus is 36.5 light-years). But another half-million years from now it will have gone much farther south in our sky and have receded back almost out of naked-eye visibility.

Spica is the second brightest star among the spring constellations. You can find it by taking the "arc to Arcturus" from the Big Dipper's handle, then driving more or less southward a "spike to Spica"—in other words, a straighter line to this very slightly blue-white *lucida* (brightest star in a constellation) of Virgo the Virgin. The name *Spica* means "ear of wheat," for it represents one being held by Virgo (a winged figure associated with the Greek Demeter or Roman Ceres, the goddess of the growing vegetation and harvest).

Regulus is the least bright of the 1st-magnitude stars, but it holds a position in the sky that makes it very special and famous. First, the slightly blue-white Regulus marks the heart of the noble Leo the Lion. Second, Regulus is the 1st-magnitude star closest to the **ecliptic**, the center line of the zodiac. This means that Regulus is frequently passed closely by the Sun, the Moon, and the planets. The name *Regulus,* which legend says Copernicus made popular, means "the little king."

The Brightest Stars of Summer

We have already met three of the four 1st-magnitude stars of summer in Sight 10; they are the stars of the famous asterism called the Summer Triangle.

The brightest star of the Summer Triangle and anywhere in the summer constellations is Vega. This decidedly blue-tinted star shines at zero magnitude, a bit brighter than Capella and Rigel and a bit dimmer than Arcturus. Only Sirius is greatly brighter, but a friend of mine calls Vega "the Sirius of Summer." Whereas Sirius is visible almost directly "behind" us—that is, in the direction from which the solar system has come, Vega is almost directly "ahead" of us—that is, in the direction in which the solar system is headed. It will take our Sun and its planets about half a million years to travel the 25 light-years that now separate us from Vega—by which time Vega itself will have moved onward. A few hundred thousand years in the future, Vega will actually replace Sirius as the star of greatest apparent brightness from Earth. About 13,000 years ago Vega was, and 13,000 years in the future it will be, a brilliant North Star for our planet. Vega was the first star ever photographed and the first bright star shown to be circled by a vast belt of dust. In recent years, research suggests that Vega has planets. I sometimes call Vega the "Sapphire of Summer" and the "Queen Star of Summer." As we noted in Sight 10, it passes almost exactly overhead for viewers at 40° N latitude. By the way, telescope users, Vega has a wonderful little line-of-sight companion star in our era of history (see Sight 45).

Altair is the lucida of Aquila the Eagle and is flanked on either side by stars of similar moderate brightness. The white (or ever so slightly yellow-white) Altair rotates about once every 6 hours—more than a hundred times faster than our Sun. This rapid spinning must cause the star to bulge at its equator, making it look egg-shaped to any nearby observers. At a distance of 16.7 light-years, Altair is the third closest of the 1st-magnitude stars visible from midnorthern latitudes.

Deneb is a white star whose name means "tail"—the position it marks in the lovely constellation Cygnus the Swan. It appears about a half-magnitude dimmer than Altair but is roughly a hundred times farther away. Thus, Deneb is a superluminous star, its true brightness greater than any of the other objects on the 1st-magnitude list. Deneb releases as much light in a single night as our Sun will in the entire twenty-first century!

Antares is an orange-gold star that marks the heart of the striking Scorpius the Scorpion. This red giant is now thought to be even bigger than Betelgeuse (bigger than the orbit of Mars). It is farther from us than Betelgeuse and so appears somewhat less bright. Along with Aldebaran, Regulus, and Spica, it is one of the few 1st-magnitude stars that the Moon can pass in front of (see Sight 22). Antares also has a remarkable green companion star near it (see Sight 45).

The Brightest Star of Autumn

There is only one 1st-magnitude star among the autumn constellations. It is the lonely *Fomalhaut* (pronounced *FOHM-uh-lawt*—not *FOHM-uh-loh*, as some people, perhaps influenced by French, think). Fomalhaut is even a little farther south than Antares. It has a circumstellar belt of small rocky objects and possibly planets—just like Vega. And by an odd coincidence, Fomalhaut is exactly as far from us as Vega—25 light-years away. Fomalhaut is the sole conspicuous star in its constellation, Piscis Austrinus the Southern (single, not plural) Fish.

The Other Brightest Stars

In this book, we are mostly limiting ourselves to the great sights that can be seen well from Earth's northern hemisphere. But six of the twenty-one stars of 1st magnitude are too far south to be viewed at all from 40° N. Observers in the southern parts of Florida, Texas, and California can at least glimpse, low in the south and therefore dimmed, the brilliant stars Canopus (second only to Sirius in brightness), Alpha Centauri (the closest star system beyond the family of the Sun, only 4.3 light-years away), Achernar, Beta Centauri, Acrux, and Becrux (the latter two are the brightest stars of Crux the Southern Cross).

Sight 16 ORION

There are sights in the heavens that are the most thrilling, lovely, eerie, or dreamlike. The one that is the noblest, among other things, is undoubtedly the sight of Orion the Hunter. A first sight of this figure of stars—first in your life or on any given night—can be literally breathtaking. After all, Orion is easily the brightest of all constellations. But even in a heavens that offers the Big Dipper for Northerners and the Southern Cross for Southerners, it is the most striking of all large or constellational star patterns. What pattern could grab our gaze as well as one that forms an idealized human form? This is especially true when the middle of that form is marked by the boldest of all small naked-eye star patterns: the Belt of Orion.

An Artistic Geometry of Stellar Beauty

How can I single out one sight of Orion that has moved me the most, or even a few of them? Perhaps no stellar sight, maybe not even Sirius (see Sight 14),

Orion.

so dependably sets the mind into a state of wondering admiration. There is a powerful simplicity about a view of a single bright star like Sirius. But the contemplation of Orion is richer and usually longer because there are a number of component stars and parts that make up the entirety of the Hunter. Let me try to describe and explain.

Orion contains the brightest blue giant and red giant stars in the sky: Rigel and Betelgeuse. They shine diagonally across from each other to the lower right and upper left of that most striking of all the sky's compact naked-eye asterisms: the almost perfect three-star row of Orion's Belt. While Rigel and Betelgeuse were major attractions among many in our previous chapter, Orion's Belt and the accompanying Sword get their own chapter (see Sight 28), and for viewers with optical aids, there is, in the Sword, the topic of Sight 46: the Great Orion Nebula (observable to most naked eyes only with some difficulty because it so closely surrounds a star). Orion is also the most important part (the bright center) of the Orion group of constellations (see Sight 9). So that is four of the fifty best sights, other than the one you are reading, in which Orion or its components are a key part.

There is something important to realize as you read about Orion, however. It is something that is understood, in both the fullest and yet most immediate way, when you actually see Orion. This understanding is that all the beautiful components of Orion combine so perfectly and intriguingly into a greater whole that that whole resembles a work of art.

Orion rising.

What do I mean? For one thing, you can analyze Orion like a work of art and appreciate why its composition is so powerful. After all, the two 1st-magnitude stars of Orion are diagonal to each other, and each has to its side (when Orion is high—or at any altitude if you are imagining the human form they depict) a lesser but still bright (2nd-magnitude) star: Bellatrix to the right or west of Betelgeuse and Saiph to the left or east of Rigel. Betelgeuse and Bellatrix are the shoulders of the Hunter, and Saiph and Rigel are the knees. But the rectangle, elongated north to south, that these four stars form is pierced through in its center with a short line of three equally bright, equally spaced 2nd-magnitude stars that is diagonal to the rectangle—the Belt of Orion. It is one of what I call the four centralities of Orion that the Belt lies halfway between the two stars of the shoulders and the two stars of the knees and makes a diagonal perpendicular to the other superb diagonal formed by the brilliant Betelgeuse and Rigel. I think most of us are aware of the rectangle at all times, but usually we are more consciously drawing the lines of Orion as tapering in from the shoulders and knees to form the waist of the Hunter that is indicated by the Belt. In the heavens, we are stunned by star patterns that are close to being perfectly geometrical and that seem to answer to our natural human tendency to recognize patterns everywhere. But any great work of art needs to vary in some interesting way from what would otherwise become a boring regularity. Orion has those variations. Rigel and Betelgeuse have their different colors, and Rigel is usually quite noticeably brighter than Betelgeuse. More important, the Belt slants jauntily in relation to the more nearly (but not exactly) east-to-west orientation of the shoulders and the knees and the nearly north-to-south orientation of the Sword. The modestly bright (but quite noticeable) stars of the Sword are not only not similar in brightness like those of the Belt but rather increase in brightness from north to south.

The Four Centralities of Orion

While observing Orion, there are two centralities of it you can easily see and two you can learn about and ponder.

The two you see are the way Orion is centered within the winter host of bright constellations (Sight 9) and the way the Belt is centered between the pairs of Betelgeuse-and-Bellatrix and Saiph-and-Rigel.

A centrality that amateur astronomers quickly learn about is the way Orion straddles the celestial equator. It is halfway between the north and south celestial poles; therefore, the constellation can be seen from anywhere in the world (at least part of it is visible even at Earth's North Pole and South Pole, and it passes overhead at Earth's equator). As a matter of fact, the Belt of Orion is virtually right on the celestial equator (the northernmost star of the Belt is only a third of a degree south of it). The Belt is by far the most prominent marker of that equator anywhere in the heavens. By the way, after you learn about this particular centrality of Orion, you can actually appreciate it roughly with the eye—by gauging that the Belt lies nearly 90° from the North Star, and that this is half the angular distance to the south celestial pole, which must be about 40° below the southern horizon for someone who lives at 40° N (just as the north celestial pole and North Star are 40° above the northern horizon for someone at that latitude—see Sight 8).

A final centrality of Orion is that most of its bright stars are blue giants of the Orion Association that are mostly about 1,200 to 1,500 light-years away. That means they are in about the center of our spiral arm—the Orion Arm—of the Milky Way Galaxy.

Three Sights of Orion

Enough of this analysis of why Orion is so artistically beautiful and the center of our attention, as well as of other things. Let me provide at least a few of the countless memorable sightings of Orion I have had over the course of almost forty-five years.

My first knowledgeable observation of Sirius and Rigel was through Venetian blinds and bare tree branches from my bed when I was six years old. I recall what must have been soon after this an observation of Orion in his entirety. As all the great observations of our life are, it was more than just an observation, it was an experience. In this case, the experience was walking down a rather steep little hill in our backyard to meet—away from the house and through breaks in the trees—Orion. Without intellectually believing I was literally journeying to Orion or that this constellation was some sort of actual personage, I felt as though I was in some sense leaving the world a little behind to go and visit Orion, to meet the great Hunter in wonderstruck veneration. Even then, I understood that while a metaphor is not a literal statement of fact, it can carry in it a truth that can change our lives—in this case, help commit me to a lifelong fellowship with the stars. The slope of that hill I walked down to meet Orion seemed greater then than it does now partly because I was a child. But it really has eroded to a much less steep slope. I'd like to believe, however, that the steepness of the slope of wonder I climb—whether down or up—is still as great today when I seek Orion.

A few years after this first remembered (and no doubt repeated) experience, I had another observation involving Orion that I've never forgotten.

This time, I was stepping outside at the loneliest hour of the night—about 4:00 A.M.—around October 20 to try to observe the Orionid meteor shower, which peaks around that date and hour. My memory is of a wondrous multitude of stars but especially of Orion hanging at his highest in the south in the most awesome stillness and silence of the stars.

Flash ahead to a November morning twilight and another meteor shower at least thirty-five years later. Dawn was bright enough to see a little color in the landscape and, gazing beyond a young fellow astronomer, I saw Orion, still dimly visible in the bluing sky. There were four or five meteors at once raining down around and through him about 30 minutes after rates in nearly full darkness had reached a thousand Leonid meteors per hour.

The Rising of Orion

I speak of Orion at his highest in the south on a winter evening and slanting down to his setting at dawn in late November. And this brings me to a final set of aspects of seeing Orion that is vitally important and wonderful. I am referring to the orientations of the valiant figure at rising, culmination (the highest point above the horizon), and setting. Far more than with any other constellation, you can't help but be aware of Orion's crossing of the sky as being a dramatic expedition. I say this because when Orion rises, the stars of the Belt form an amazing vertical line, and because his form is tilted at an angle that suggests the human figure, his stars represent leaping up steeply from the horizon. One minute there is only anticipation. Then there is a glimmering of a few stars in the horizon mists. Then, sometimes with what seems like stunning rapidity, the bright form of Orion—looming huge in the same way a rising Full Moon does—flashes into full visibility. This experience is incredibly impressive, so uplifting of the spirit with the uplifting of Orion. In the novel *The Lord of the Rings*, J. R. R. Tolkien describes this sight as being met with shouts of gladness and songs of joy by the company of elves who have just rescued Frodo and his hobbit friends and are now celebrating the rising of the figure they call Menelvagor or Menelmacar—the Swordsman of the Sky.

Writers sometimes talk about Orion "striding" across the sky as the night passes. How amazing it is that the constellation of stars that vaults up tilted from the horizon achieves its perfect uprightness when it gets to the highest point in its arc across the sky. Orion stands straight up in strength in the south on the north-south meridian of the sky, the halfway point of the journey across the heavens.

There is one final thought I always have about Orion. It is a call to a new mythology about this constellation. The character this constellation is named after was, in Greek mythology, gigantic and the "handsomest of men"—though really half-god, because his father was Poseidon (the Roman deity Neptune), god of the sea. The ancient Greeks said that Orion was either so tall that his head was above water level as he walked along the sea bottom or that he gained from his sea-god father the ability to walk across the waters. All of this sounds fittingly wonderful and glorious for a character who became associated with the noblest and brightest of the constellations. But the Greek character Orion was anything but noble in his vain, aggressive, and lustful behavior. I have always disagreed with this. To me, the splendid pattern of stars we call Orion should represent for our future a different sort of hunter from this one who bragged that he could kill any animal. Let's have the constellation Orion represent the inquisitive spirit of humankind and all that is best of it. Let Orion represent the part of us that longs to explore and to hunt for the noblest of quarry—the truth.

Sight 17 Other Prominent Constellations

Although Orion stands out as the most spectacular of the eighty-eight official constellations, there are many others that are among the worthiest sights in astronomy. In this chapter, I will try to touch on almost all the constellations easily visible from midnorthern latitudes. Many are not very bright (to see them, you will need dark skies for your naked eyes or, if you are at a light-polluted site, binoculars). But all are, in one way or another, interesting. To locate these constellations, refer back to Sight 1 and the figures in that chapter.

Winter Constellations

Winter is the season when other constellations have to compete directly with Orion (see Sight 16). None are as bright or as attention-grabbing in their shape as Orion, but each is an individual, and several have forms and component stars that are fascinating in their own right.

Taurus the Bull features the bright and orange Aldebaran, which with the giant Hyades star cluster forms the V-shaped pattern of stars that depicts the face of the Bull (for a close-up on the Hyades, see Sight 25). Taurus also contains the loveliest and most remarkable of all naked-eye star clusters: the Pleiades (see Sight 26), which are located on the imagined shoulder of the Bull. Taurus is one of the twelve constellations of the zodiac, so the Moon and planets make many impressive journeys through it. The rest of Taurus is marked by two stars that represent the tips of the Bull's rather long horns. The brighter of the two stars, Beta Tauri, links the pattern of Taurus to that of Auriga the Charioteer. The dimmer of the two horn stars, Zeta Tauri, assists in locating one of the most fascinating of all telescopic objects: the supernova remnant M1, also known as the Crab Nebula (one of the objects mentioned in Sight 30). Taurus gains some of its interest by seeming to face and challenge the mighty Orion.

Auriga the Charioteer uses Beta Tauri to form a fairly prominent pentagon of stars. Its brilliant star is Capella, which is accompanied by the little triangle of "the Kids." Auriga is also home to three fine open clusters for telescopes: M36, M37, and M38 (see Sight 47).

Gemini the Twins is the next constellation of the zodiac after Taurus. It earns its name from virtue of the fact that its two brightest, and similarly bright, stars are located only about 4½° from each other. These two, the 1st-magnitude Pollux and almost 1st-magnitude Castor, are named for the most famous twins of Greek legend. The rest of Gemini includes the 2nd-magnitude star Alhena (also called Almeisan), which lies in the southern feet of the Twins. Gemini also contains, near the northern feet, the star cluster M35 (see Sight 47), which is dimly visible to the naked eye in dark skies. Additionally, Gemini offers two very different naked-eye variable stars (see Sight 29).

Canis Major, the Big Dog, is best known for its head or heart, which is the brightest of all stars, Sirius (see Sight 14). But Canis Major is also one of the few most brilliant constellations in all the heavens. Its star, Adhara, is the brightest of all 2nd-magnitude stars, slightly outshining the much more famous Castor. The rather compact triangle of stars that forms the southern half of Canis Major is almost as bright as Orion's Belt, although far enough south to be behind trees or buildings or dimmed by haze for many Northerners. Just a few degrees south of Sirius itself is the star cluster M41 (see Sight 47), which is bright enough to be glimpsed with the unaided eye in dark skies. To the right (west) of Canis Major and under the feet of Orion is Lepus the Hare, which is brighter than people think but suffers by comparison to the resplendent Orion and Canis Major.

Canis Minor, the Little Dog, contains the bright star Procyon and not much else—the little pattern looks more like a tail than a dog. Between Canis Minor and Canis Major is a very dim naked-eye constellation with some superb telescopic sights: Monoceros the Unicorn.

Spring Constellations

Leo the Lion announces spring by climbing up the southeastern sky on March evenings. The pattern of Leo looks very much like the outline of a lion. A backward question mark of stars represents the noble head, the mane, and the chest of the royal beast, and a right triangle of stars forms the hindquarters a considerable distance to the left (east) of this front part. The backward question mark is also pictured as being a westward-facing hook called "the Sickle." Leo is a constellation of the zodiac and its brightest star, Regulus, is the 1st-magnitude star closest to the ecliptic and therefore a frequent participant in conjunctions with the Moon and the planets. Regulus marks the heart of the Lion, the dot of the backward question mark, or the handle-end of the Sickle. Midway in the hook of the Sickle is the 2nd-magnitude Algieba, a fine telescopic double star (see Sight 45). The easternmost or hindmost portion of Leo is 2nd-magnitude Denebola, said to mark the tail of the Lion, even though in ancient times it was the Coma Star Cluster (see Sight 27), located farther to the northeast, that formed the tuft of the tail.

Hydra the Sea Serpent, is the longest of all constellations in the east-west direction. It's only fairly conspicuous parts are its modestly bright but compact head and its lonely 2nd-magnitude orange heart, Alphard.

Ursa Minor the Little Bear, the bearer of the North Star, is the constellation that contains the Little Dipper (see Sight 8). Ursa Major the Great Bear, contains the Big Dipper (also profiled in Sight 8), but the rest of the constellation is interesting to trace, high in the north on spring evenings. And right under the handle of the Big Dipper is the dim Canes Ventici, the Hunting Dogs, which contains an important galaxy (see Sight 50). The Big Dipper's handle curves out to point to the brilliant Arcturus in its fairly dim but kite-shaped constellation Bootes the Herdsman, which is home to many telescopic double stars (see Sight 45). Corvus the Crow, contains no really bright stars but is rather conspicuous because it forms a compact rhomboid figure not far from Virgo's 1st-magnitude Spica.

Virgo the Virgin and Cancer the Crab are dim constellations overall but members of the zodiac. Virgo is long and rather amorphous, but it contains Spica and a wonderland of telescopic galaxies (see Sight 34). This observer's paradise of many galaxies spills over into the faint Coma Berenices (Berenice's Hair), named for the big, disheveled Coma Star Cluster (see Sight 27), which can be seen by the naked eye. Another dim zodiac constellation of late spring or early summer is Libra the Scales, which contains an easily split double star (see Sight 45).

Constellations of Summer

We discussed Lyra the Lyre, Cygnus the Swan, and Aquila the Eagle as the three best constellations of the Summer Triangle in Sight 10. Also in and around the Summer Triangle are Delphinus the Dolphin, Sagitta the Arrow, Vulpecula the Little Fox, and Scutum the Shield. Even though they are all dim constellations, they contain several superb telescopic objects (read about Scutum's open cluster M11 in Sight 47, Vulpecula's planetary nebula M27 in Sight 49, and the naked-eye Milky Way's Scutum Star Cloud in Sight 7).

Scorpius the Scorpion is the second brightest of all constellations (after Orion), but it shines on summer evenings when Orion is not visible. Scorpius is a constellation of the zodiac, although the ecliptic actually only passes through a fairly short northwestern part of the constellation. The heart of Scorpius is marked by the 1st-magnitude Antares, the brightest red giant of summer. But Scorpius doesn't depend on Antares to be prominent. Its pattern is striking—a kind of double twist or fallen-forward letter S that does in fact suggest the shape of a coiling scorpion. This pattern contains more stars brighter than magnitude 3.0 than any other constellation.

Well to the right (west) of Antares is the head of the Scorpion, a more or less north-south line of fairly bright stars. In this line are Beta Scorpii (Graffias), a fine telescopic double star (see Sight 45) that is so near the ecliptic that the planet Jupiter passed in front of it once back in the early 1970s. Also in this "fence of stars" (as my friend Guy Ottewell calls it) for Moon and planets to pass through is Delta Scorpii (Dschubba), a star that seemed rather normal until a few years ago, when it suddenly brightened considerably. It became as bright as magnitude 1.6, almost reaching 1st magnitude, and fluctuated in brightness (as it was still doing at the time of this book's writing). Antares itself is flanked to either side by moderately bright stars (somewhat like Altair), which are together known as the Praecordia, meaning the "outworks of the heart."

Just a few degrees west of Antares is the big, close (but mostly telescopic) globular cluster M4 (see Sight 48). The rear half of the Scorpion curves steeply south from Antares, and this emphasizes a problem for northerly observers who want to enjoy Scorpius: if you live farther north than about 45° N, the very bottom of Scorpius never gets above your due south horizon. Even observers at 35° N wish that Scorpius were higher in their sky, especially so that it wouldn't be so greatly dimmed by summer haze. The tail does curve back up northward some distance, where it is tipped with "the sting": magnitude 1.6 Lambda Scorpii (Shaula) and magnitude 2.7 Upsilon Scorpii

(Lesath)—two stars so close together that to the naked eye they form a striking pair, sometimes called the "Cat's Eyes" (though one eye is markedly brighter). To the upper left from the Cat's Eyes is a pair of big star clusters bright enough to be seen as fuzzy patches of light with the naked eye under very good sky conditions. These are the open clusters M6 and M7, which are part of the incredible richness of the Sagittarius region in the direction of the center of the Milky Way Galaxy (see Sight 32).

Sagittarius the Archer is marked by a quite noticeable pattern of fairly bright stars (several are 2nd magnitude) called the "Teapot." But in dark, clear skies it is the seeming "steam" of the brightest Milky Way (see Sight 7) and the many "deep-sky objects" that can be seen with the naked eye, binoculars, and a telescope in this part of the heavens (again, see Sight 32) that are the most thrilling.

High in the summer sky between the brightest spring star Arcturus and the brightest summer star Vega shine two fairly faint but interesting constellations: Corona Borealis the Northern Crown, and Hercules the Strongman. The former is a semicircle of stars with one 2nd-magnitude star called Gemma or Alphecca and several fascinating (though usually not very bright) variable stars (see Sight 45). Hercules is composed partly of a geometric pattern of stars called the Keystone, in which the famous globular cluster M13 (see Sight 48) is located. A third fairly dim summer constellation very much worth observing is Draco the Dragon. It curls in the northern sky mostly between the Big Dipper and the Little Dipper, but its compact head, which includes the 2nd-magnitude Eltanin, points toward Vega. Midway up the southern sky, between high Hercules and low Scorpius, are the big and sprawling Ophiuchus the Serpent Bearer (mostly dim despite the 2nd-magnitude Ras Alhague) and Serpens the Serpent—the only constellation divided into two separate parts (by Ophiuchus—the western part is Serpens Caput, the Serpent's Head, and the eastern part Serpens Cauda, the Serpent's Tail).

Autumn Constellations

Autumn's bright constellations pass high in the north. A vast area of the southern and southeastern skies on autumn evenings is filled with dim constellations that all have a connection to water. Therefore, this region is sometimes called the "Water" (or, much less often, the "Great Celestial Sea"). It includes three constellations in a row of the zodiac that are dim: Capricornus the Sea Goat (front-half goat, back-half fish), Aquarius the Water Bearer, and Pisces the (two) Fish. All these have a few sights of interest for naked-eye observers with very dark skies but especially for telescopic observers (for three marvelous telescopic objects in Aquarius, see Sights 48 and 49). Also in the Water are Piscis Austrinus the Southern (one) Fish (which has one great attraction: autumn's sole 1st-magnitude star, Fomalhaut) and Cetus the Whale (which has two 2nd-magnitude stars on either end of its long dim body but not much else, except for the amazing variable star Mira—see Sight 29).

Pegasus the Winged Horse represents the famous creature from Greek mythology and has a 2nd-magnitude star (Enif) for a nose, sniffing for the mostly telescopic globular cluster M15 (see Sight 48). But the most important part of the constellation is the fairly prominent and very useful asterism called the "Great Square of Pegasus." The Great Square consists of four stars that range from magnitude 2.1 to 2.8, though the brightest of the four technically belongs to the neighbor constellation Andromeda. A line extended south down the western side of the Square eventually brings the eye near the very southerly Fomalhaut. A line extended south from the eastern side of the Square finds Cetus's tail star, Diphda (also called Deneb Kaitos), and extended north finds the western end of Cassiopeia.

Cassiopeia the Queen is a zigzag of stars that range from magnitude 2.1 to 3.4 and hang high in the north on autumn evenings. Telescopes show many beautiful star clusters in Cassiopeia. Not far west of Cassiopeia in the high northern sky is her mate, dim Cepheus the King with one of the sky's most important variable stars (see Sight 29). Cassiopeia and Cepheus are named for characters in one of the truly great Greek myths. This myth is represented by six constellations and covers 12 percent of the entire celestial sphere and about one-fifth of the entire sky when they are all above the horizon. The six constellations are Pegasus; Cetus (in the original story not a whale but a sea monster of unspecified type); Queen Cassiopeia and King Cepheus and their daughter, Andromeda; and the myth's central person—Andromeda's rescuer, Perseus.

Perseus shines like a bright letter K of stars, climbing the northeastern sky on autumn evenings. Perseus normally sports two 2nd-magnitude stars, Mirfak and Algol, but the latter undergoes spectacular dimmings about every third day as the sky's most prominent **eclipsing binary** (see Sight 29). Perseus offers several very bright and marvelous star clusters, all capable of being seen with the naked eye, and all much more spectacular with low-magnification optical aid: the Alpha Persei (Mirfak) Cluster, M34, and the telescopically breathtaking Double Cluster of Perseus (see Sights 27 and 47).

Stretched between Algol and Pegasus is a constellation as lovely as the princess it is supposed to represent: Andromeda the Chained Maiden. Andromeda was chained to a sea cliff for sacrifice to the sea monster Cetus in order to propitiate the gods and keep them from sending the monster to ravage the people (the gods were angry because Cassiopeia had claimed her own beauty was greater than that of the sea nymphs). Perseus swooped in on his winged sandals (or, some storytellers say, on the winged horse Pegasus) to save Andromeda. He did this by using the head of the hideous snake-haired Gorgon sister Medusa to turn the sea monster to stone (whoever or whatever saw the face of Medusa—even after Perseus decapitated her—was literally petrified, or turned to stone). As a constellation, Andromeda is dominated by a long line of three stars, all the same brightness and equally spaced from one another. But Andromeda is best known in modern times for its galaxy: M31, the Andromeda Galaxy, visible even to the naked eye as a longish smear of fuzzy light (see Sight 33).

Aries the Ram is a zodiac constellation represented mostly by a tiny bent line of three bright stars, one of which is bright—2nd-magnitude orange Hamal. Between Aries and Andromeda is the also little but entirely dim constellation Triangulum the Triangle, which is noted for its big but dimmer cousin of M31, the face-on spiral galaxy M33 (see Sight 50). M33 is just visible to the naked eye under very dark sky conditions.

East from Aries, pointed at by one of the legs of Perseus, shines the utterly beautiful Pleiades star cluster—so we have returned to Taurus and completed a full year of the constellations in this tour.

Field of View

15° to 1°

(Narrow Naked-Eye Field, Binoculars Field, and Wide-Telescopic Field)

TOTAL ECLIPSE OF THE MOON

Sight 18

Everyone can understand that an event that greatly dims and vividly colors the Moon would be lovely and fascinating. But something else that has always impressed me about total lunar eclipses is the way they help restore the wonder we should never cease to have when we look at the Moon—eclipsed or not. If you've gazed at the Moon with the naked eye so many times that you've grown complacent about it, a total eclipse of the Moon will change that. If you've observed the craters, mountains, and other features on the Moon through a telescope enough times for the edge of amazement to be dulled a bit, a total eclipse of the Moon will change that, too.

Total lunar eclipses are underrated. They are often described together with total solar eclipses and suffer by comparison. But what celestial sight doesn't? Total eclipses of the Sun are so staggeringly beautiful (and beautifully staggering) that you might forget your name, even your world, for the few minutes one is going on. But total solar eclipses last only a few minutes, and unless you have the freedom and finances to globe-trot, you will probably get a good chance to see only a few of these events in your lifetime. The reason most of the human race will never see a total solar eclipse is the extremely limited area in which such an eclipse is visible. The exact opposite is true of total eclipses of the Moon. They actually occur less frequently than total solar eclipses. But when a total lunar eclipse does occur, it is visible from slightly more than half of the Earth—the half that is experiencing night during the total eclipse. And the total stage of a lunar eclipse does not typically last for 3 or 4 minutes (like a solar totality) but 40 or 80 or even more minutes.

What's more, there is much of interest to see during the entire several hours of a lunar eclipse, at least during the entire partial and total stages. There is somewhat less to see (and the Sun itself only with special filters or projection) during the partial stages of a solar eclipse.

The bottom line is that a total eclipse of the Moon is a leisurely but suspenseful several hours of slowly but steadily transforming loveliness and celestial interrelationship. And it is this—an exercise in extraordinary evolving heavenly beauty—a number of times each decade, an exercise during which you know that you are witnessing drama and beauty shared with an entire hemisphere of our planet.

A total eclipse of the Moon.

The Early Stage

I was originally going to relegate our discussion of the nontotal stages of lunar eclipses to Sight 37, which does deal with partial solar eclipses. But I only *mention* partial lunar eclipses there now. Partial solar eclipses, you see, are so much milder and different from total solar eclipses and are almost always seen by an observer at a given location on Earth without culminating in totality. But about half of all lunar eclipses in which Earth's umbral shadow touches the Moon's face for an observer will culminate in totality. Furthermore, a real taste of some aspects of a total lunar eclipse can be had by just watching a very large partial one. That's not at all true of a solar eclipse, for it is the last fraction of 1 percent of the Sun being hidden that reveals virtually all the greatest wonders of totality.

So let's see how a lunar eclipse progresses from its first trace up to the moment that the total eclipse begins.

Interestingly, unlike a solar eclipse, a lunar eclipse has a stage before partial. This stage is the *penumbral eclipse.* An eclipse of the Moon occurs when the Moon passes into Earth's shadow. But there are two fundamentally different parts of Earth's shadow, or of all shadows for that matter: the penumbra and the umbra. The *penumbra* is lighter than the umbra and is peripheral. It exists where only part of the Sun is hidden by Earth. In other words, if we were standing on the Moon during a lunar eclipse, we would first have only part of the Sun covered by Earth—which would dim the landscape around us, but not greatly. Meanwhile, a viewer on Earth would not even be able to tell that the penumbra was starting to move across the Moon until the penumbra was about halfway across the Moon. At that point, the Moon's edge that was deepest into the penumbra would finally start being deeply dimmed enough for us to notice it as a kind of slight stain.

The *partial eclipse* starts when the Moon has completely entered the penumbra and then first encounters Earth's central shadow—the *umbra.* A viewer on the part of the Moon where the umbra was now touching would see the Sun entirely hidden by the giant black bulk of Earth. But the rest of the Moon would still be only slightly dimmed by the penumbra.

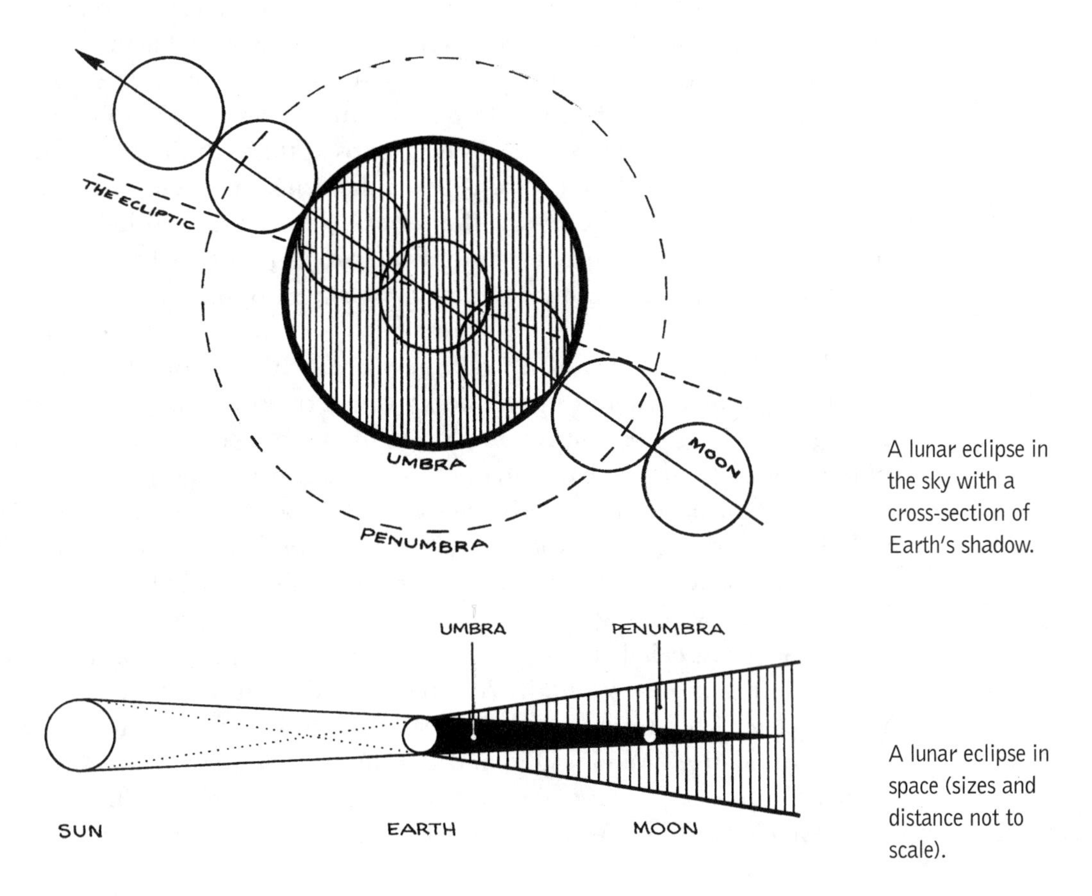

A lunar eclipse in the sky with a cross-section of Earth's shadow.

A lunar eclipse in space (sizes and distance not to scale).

Are you having trouble picturing the relation of the penumbra to the umbra? Then just take a look at the figures on page 103. You can also study the penumbra and umbra of your finger's shadow or those of a little ball held above a white sheet of paper underneath a desk lamp.

The Partial Eclipse Advances

Whereas the first touch of the penumbra on the Moon's face was too slight to even be perceivable, the first touch of the umbra is unmistakable—almost instantly apparent in a telescope and quickly apparent to the naked eye. The partial eclipse has now begun, and the umbra seems at first almost inky dark.

Watch the umbra start creeping across the lunar landscape. One wonderful lunar feature after another is reached by the wave of darkness and slowly overwhelmed by it. Once the umbra has moved far enough onto the Moon for the naked eye to make out its shape, we are treated to a thrilling direct demonstration of an important fact about our world. The edge of the umbra is curved. This is visible proof that our planet is round. Indeed, lunar eclipses were the first indicator to people in ancient times that the Earth is round.

If you are watching the progress of the umbra through a telescope, you soon notice that it is not completely dark. Lunar craters and mountains—or at least the brighter ones—can still be glimpsed through the veiling umbra. In fact, depending on how dark this particular eclipse is, a telescope may start showing you a hint of redness in the umbra before it is even halfway across the lunar disk. By the time the umbra is most of the way across the Moon's face, even the naked eye may be able to start glimpsing red in parts of the umbra.

Why red? Why any color or brightness in Earth's shadow? Before we discuss the absolutely astounding answers to these questions, make sure you are noting something else that is going on as the umbra covers more and more of the Moon: the stars are coming out as if a deeper night were falling within night. Of course, you need to be at a location that doesn't have much light pollution to witness the glory of this coming out of the stars. When the eclipse starts, the sky is soaked with bright moonlight—the light of the Full Moon—and few stars are readily visible. But while you are staring at the Moon itself—especially if you're watching this event through a telescope—the darkening of the sky and the coming out of the stars can sneak up on you. You step back from your eyepiece and suddenly notice that many times more stars than before have emerged and that bright ones have taken on prominence. For any skywatcher who loves the stars and the night, this side effect of a lunar eclipse is a beautiful and gladdening one. And you hold your breath with anticipation, wondering: how much darker will the sky get and how many more crowds of stars will appear, when the Moon is completely covered by the umbra?

All the World's Sunrises and Sunsets at Once

Now we turn to the question of why we are starting to see a hint of red in the umbra. The red is still suppressed by our eyes being overwhelmed by the brightish part of the Moon outside the umbra. But as total eclipse nears and arrives, we are probably going to see the red become dominant. It's not a fact that everyone knows, but fact it is: when the Moon is totally eclipsed, it rarely just blacks out; it usually turns some vivid shade of red or orange. Why wouldn't Earth's umbra be completely colorless and dark?

The reason is that Earth's atmosphere *refracts*—that is, bends—the light from all our world's sunrises and sunsets at once into the umbra. We've all seen that when the Sun is low in the sky it is reddened by passing through a long section of air and haze. In the case of a lunar eclipse, the sunlight continues onward, taking an even longer grazing path along the curve of the Earth, so it is even more reddened. What is cast on the Moon is not so much a dark shadow as a gloriously tinted beam of light. It is almost like a colored spotlight being cast onto a stage performer—but this "performer" is the Moon. Think, too, of what a total lunar eclipse would look like from the Moon: the black shape of Earth would pass in front of the Sun and be surrounded by a narrow ring of reddish light.

I keep saying reddish but in reality what makes total lunar eclipses even more suspenseful is that the color and brightness of the Moon can be bright and orange, moderately dark and coppery, quite dark and blood red, or—in very rare cases—completely gray or black. The determining factor will be how clear the atmosphere is in that Earth-encircling band where the sunrises and sunsets are occurring at the time of the eclipse. If there happens to be a lot of clouds in the band, the eclipse will be darker than average. But the only thing that can make the Moon's face so dark that it just about becomes invisible is if Earth's atmosphere higher up is blocked by vast areas of ash or sulfuric acid haze produced by a great volcanic eruption. During much of the twentieth century, there was no eruption that produced a really dark Moon. But at the eclipses of December 30, 1963, and December 30, 1982, which occurred soon after the eruptions of Mount Agung and El Chichon, respectively, the Moon was rendered almost invisible. The total lunar eclipse of July 6, 1982, occurred when El Chichon's sulfuric acid cloud had not yet spread very far north so that the Moon appeared pied—partly black and partly red.

Details on these dark lunar eclipses and many of the joys of watching lunar eclipses are discussed in my book *The Starry Room* (see the sources section in the back of the book).

There is a special scale for rating the color and darkness of a total lunar eclipse: the Danjon Scale (see the table below). How bright, in terms of magnitude, is the Moon at the different levels of the Danjon Scale? You may be surprised at how much the Moon is dimmed. My own study of this over the years indicates that a Danjon rating of 3.0 equals a magnitude of about –3.0 (a little brighter than Jupiter), and that a Danjon rating of 2.0 equals a magnitude of about 0.0 (similar to a star like Vega, Arcturus, or Capella). The Moon during the very dark eclipse of December 30, 1963, received an average Danjon rating of just 0.2 from observers, who estimated the magnitude of the Moon as only about 4.1. If you don't think these magnitudes—or at least the –3.0 and 0.0 magnitudes—are dim for the Moon, consider two things. First, the normal magnitude for the Full Moon is about –13—about 10,000 times brighter than the eclipsed Moon at a Danjon rating of 3.0. Second, remember that this is the total light of the Moon and it is spread out over a much larger area than the light of a planet or a star. The moon at a rather dark eclipse is an eerie or even ominous dull blur of deep red or red-black.

The Danjon Scale of Lunar Eclipse Brightness

L stands for luminosity.

L = 0	Very dark eclipse; Moon hardly visible, especially near mid-totality
L = 1	Dark eclipse; gray to brown coloring; details on the disk hardly discernible
L = 2	Dark red or rust-colored eclipse with a dark area in the center of the shadow, the edge brighter
L = 3	Brick-red eclipse, the shadow often bordered with a brighter yellow edge
L = 4	Orange or copper-colored, very bright eclipse with bluish bright edge

There are several ways to estimate the Moon's magnitude during an eclipse. If you are nearsighted and wear glasses, you can take off your glasses and compare the blur of the Moon to the greatly out-of-focus images of planets and stars (don't forget to use a comparison object that is roughly as high in the sky as the Moon or to correct for "atmospheric extinction" if either the Moon or the comparison object is low in the sky). A more sophisticated method involves looking at the Moon through binoculars backward to reduce the size of its image and comparing that image to the naked-eye view of planets and stars. When you do this, you must apply a correction factor (which depends on the size of your binoculars) to get an accurate result. See Philip S. Harrington's book *Eclipse!* (see the sources section) for more details on this method. As an alternative (or complement) to the Danjon Scale, you might consult the Fisher Scale. The Fisher Scale rates the darkness of the eclipse on how much surface detail you can see on the eclipsed Moon's face with the naked eye, binoculars, and a telescope.

The Endless Variety of Lunar Eclipses

The descriptions offered by the Danjon Scale suggest that there are sometimes colors other than the predominant shade of reddishness on the Moon's face. Technically, we should think of these colors as being in different parts of the umbra and merely being intercepted by different parts of the Moon during the eclipse. The variety of different shades of color seen with the naked eye but especially with optical aid in parts of the umbra is tremendous. Over the years, I have seen almost any color you could name in a lunar eclipse at one time or another, even traces of purple, visible to the naked eye!

What determines how dark a lunar eclipse will be is not just the cloudiness or volcanic material in Earth's sunrise-sunset band but also how deeply into the umbra the Moon passes. The more central the Moon's path through the umbra, the darker it will tend to be (all else being equal—which it may not be). The centrality of the Moon's passage through the umbra also, of course, helps determine how long the total eclipse lasts. Some are 30-minute jaunts through the outer part of the umbra. Others can last up to about 106 minutes if the Moon goes through the center of the umbra and is farther than average from Earth so that the Moon is small and traveling more slowly than usual.

Another unknown when you watch a total lunar eclipse is how burstingly bright the first section to escape the umbra will be. During some eclipses, the emergence of the Moon from the umbra is so dramatic that it is almost like a spotlight being turned onto one's upturned face over the course of just two or three minutes. This return of light tends to have an uplifting, cheering effect on the observer. But it may be needed to counteract the melancholy of knowing that the main event is over. One always wishes for more, and there are spells when no total lunar eclipse is visible anywhere in the world for several years. But the feeling that typically wins out at the end of a total lunar eclipse, I think, is a deep satisfaction at having spent so much enchanted time enjoying a long banquet of colors, brightness, and vast yet precise motions and positionings in the heavens.

Sight 19

TOTAL ECLIPSE OF THE SUN CLOSE-UP

In Sight 2, we took in the overall experience and sight of the most awesome natural event that humankind is capable of predicting: a total eclipse of the Sun. That event is more than enough to weaken your knees with wonder even if you witness it without the help of any optical aid. But if you do have binoculars and a telescope then there will be some extra lightning of wonder shooting through your nerves. You will get to see intricate and colorful details of structures that otherwise cannot be glimpsed directly from the planet Earth—or from anywhere else other than the shadow of the Moon.

The Inner Corona and the Prominences

Although it's safe to view the diamond-ring effect with the naked eye without protection, or even Baily's Beads, you do need to be careful. To look at these marvelous phenomena with binoculars or a telescope, however, is too dangerous for a beginner—or, to my mind, even for an eclipse veteran. Enjoy them with a naked-eye glimpse. The stage to enjoy safely with an unfiltered telescope or binoculars is the total eclipse when the Sun's blinding surface is 100 percent covered by the Moon.

The dark silhouette of the Moon looms large in your binoculars or low-magnification telescopic eyepiece. And what absolutely transfixes you is the exquisite intricacy of the structures in the "crown" of the Sun—the corona. The overall shape of the prominent part of the corona is different at each eclipse but characteristically different when the Sun is near solar maximum versus solar minimum. When the Sun is near maximum in the roughly eleven-year solar cycle of activity, its corona is brighter and possibly larger, but it is also more symmetrical in shape. When the Sun is near solar minimum, the corona is more irregular and tends to consist of large individual "petals" that appear to some observers to resemble a dahlia blossom. Around solar minimum, there are likely to be beautiful wide "polar brushes" extending out from the top and bottom of the Sun. Whatever the point in the solar cycle, you may observe a marvelous variety of structures in the inner corona—including what seem to be vast loops in the pearly light.

Now, your initial awe at even the bright structures of the overwhelmingly beautiful corona may soon be distracted by another striking sight: tufts of

The corona at the 1991 total eclipse of the Sun.

bright red sticking out from the edge of the silhouetted Moon. These are *prominences*—enormous fountains of solar gas that are structured by magnetic field lines and are seen during the eclipse frozen in place by distance and immensity. These colorful appendages (though they may actually float well above the hidden blinding surface, the photosphere) are sometimes tens of thousands of miles long or more, dwarfing the Earth. So they are big enough to sometimes be glimpsed with the naked eye during totality. One of the things you can appreciate properly with optical aid, however, is how the dark shape of the Moon creeps along to cover (or uncover) them. Prominences are most likely to be seen near the start or end of totality but sometimes a large or a high one remains visible throughout the eclipse if the Moon is moving perpendicularly to its position.

The Outer Corona and Final Overview

The final precious seconds of totality are slipping away. But now that your eyes have begun adjusting to the reduced lighting levels, you may wish to see how far out you can trace the outer corona. Use binoculars to try tracing the longest streamers of the corona. Sometimes such streamers can be followed for several degrees.

You will also want to scan—probably with binoculars—the overall view of the corona from its brightest rim around the Moon to its most delicate far-reaching wisps. The eye excels at seeing detail over a tremendous range of brightness. No ordinary photograph has ever been able to capture details of the outer corona without overexposing the inner corona. Only since the early 1990s has digital technology made it possible to combine different exposures and create a picture that portrays a view similar to what the eye sees.

You have tried your best to keep track of the time during the total eclipse. It's difficult to remember the time or your plans, even if you repeatedly rehearsed them, when you are being thunderstruck by the reality of a total eclipse of the Sun. But you can heed the warning signs of the end. If the Moon's shadow is plainly visible, its rear edge is nearing the Sun. If new prominences became visible at the rear edge of the Moon, they are now being exposed more and more. Pull back from your unfiltered telescope. It is time to watch for a farewell diamond ring and take in the entire scene as the sky's most awesome few minutes come to a stunning conclusion.

Note that the partial stages of a total solar eclipse are an element of its overall drama. You can observe these stages with special filters or techniques (see Sight 37).

THE MOON AT FULL AND OTHER PHASES

Sight 20

If we weren't so accustomed to seeing the Moon, we would realize that it is the most stupendous wonder of the night sky, more stunning than the rarest bright comet. As it is, there is a way we can regain some of our native wonder for Earth's sole natural satellite even when there isn't an eclipse. We can do it by focusing on certain aspects of the Moon that stand out when it is at various phases and positions in the sky. We will leave to the next chapter the special sights that can be presented by the Moon on the relatively rare occasions when it is seen as a very thin crescent. But what about Full Moon, the Half Moon, and the Gibbous Moon?

The Full Moon, the Rising Moon, and the Moon Illusion

The diagram on page 113 explains why the Moon shows us different phases at different positions in its orbit. If you know what the phase of the Moon is, you should also be able to figure out where and when it is in different parts of the sky, or vice versa (if you know the time of night it will be in a certain direction, you can figure out what its phase will be). But Full Moon is one of the easiest phases to understand. Two chapters back, we saw that lunar eclipses occur at Full Moon and that Full Moon occurs when there is approximately a straight line of Sun-Earth-Moon in space. When this happens, all the parts of the Moon facing the Sun (and hence lit up, experiencing day) are also facing Earth. It follows that the Full Moon also appears in the opposite direction from the Sun in our sky: so it rises around sunset, is highest in the middle of the night, and sets around sunrise.

The Moon appears almost perfectly round for several days around Full Moon and the completeness and roundness of that orb is impressive, indeed seemingly capable of evoking strong human sentiments of varied kinds. The emotions seem to range from those of the romantic to those of the supposed lunatic or even (believers in the occult tell us) the lycanthrope. Of course, it's not just the roundness of the Full Moon but the relative strength of its light that is remarkable. While the Moon seems to be round for several nights, its surface reflects a lot more sunlight to us on the exact night of the Full Moon.

The Full Moon.

You might think that the Moon shines half as bright as the Full Moon when it is half lit—which happens about seven days before the Full Moon. Actually, the half-as-bright time comes when the Moon is only about two and a half days before full.

The Full or nearly Full Moon is the one people most often notice rising (because it rises around dusk when lots of people have not yet retreated indoors for the night). The rising (or setting) Moon, especially the Full Moon rising in dusk, is famous for often appearing very much bigger than it normally does. The complete explanation of this "Moon Illusion" is not simple (it is connected with our usually unconscious perception of the shape of the sky). But an obvious factor is that when the Moon is rising we see it seemingly beside familiar objects in the landscape that are far away (if they were closer, they would be blocking our view of the rising Moon). A distant tree looks small, so if the Moon is rising on the horizon beside it, the Moon will look big by comparison (this kind of illusion is hard to get when the Moon is high in the sky because there is then seldom any distant object beside it to compare it to).

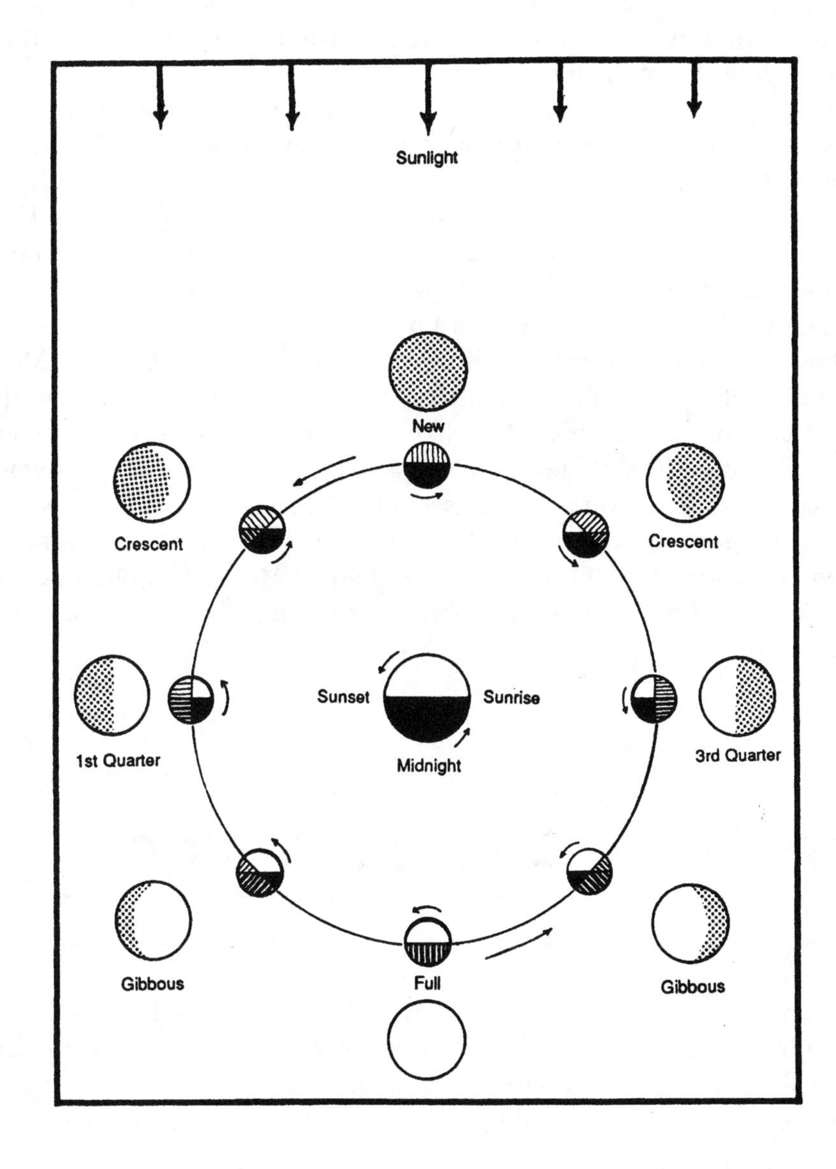

The Phases of the Moon.

Half Moons and Gibbous Moons

Whereas it is difficult to tell for sure by visual inspection of the Moon exactly which is the night of Full Moon, quite the opposite is true of a half-lit Moon. A careful observer can often estimate the time when the Moon is exactly half lit to within a few hours. The key is our built-in human ability to recognize rather precisely the straightness of a line. In this case, we are looking for the straightness of the *terminator*—the line that separates light and dark, day and night on the Moon. There will often be small irregularities in the terminator caused by differences in the height of lunar terrain (hills and highlands will catch the light of the rising Sun before valleys and lowlands do). But it is the overall straightness of the terminator you try to judge.

Many people are confused when they learn that the Moon is half lit at the phases called First Quarter and Last Quarter. Why "quarter" if the Moon is half lit? Quarter in this usage refers to the fraction of the lunar month that has transpired between the previous New Moon and the next to come. Thus, at First Quarter the half-lit Moon has completed its first quarter—its first one-fourth—of the full cycle of phases from New Moon to New Moon. At Last Quarter, the Moon is beginning its last one-fourth of that cycle.

When the phase of the Moon is larger than at the invisible New Moon phase but smaller than half lit, we call it a **crescent**. We will talk about this special phase in the next chapter. But what do you call the phase between half-lit and Full Moon? This is the **gibbous** phase. It is actually the shape we most often see the Moon in, yet it doesn't seem to be recognized and appreciated (nonastronomers' illustrations of the Moon almost always show it as full or crescent, occasionally half lit). This is the phase of Moon most often noticed in the day sky—a time when the Moon looks like a pale white piece of fine china.

Very Thin Crescent Moon

The crescent Moon has captured the imagination of humankind in wondrous ways. Some countries proudly have such a Moon on their flags. Artists may use a lunar crescent to help establish a mood of anxiety or loneliness in their work. But things with "edge" are often said to be exciting or stimulating, and the lunar crescent is all edge! The yawning incompleteness of the crescent may suggest not only emptiness but also a more desirable powerful longing. Almost inconsistent with most of these feelings—yet not so—are the feelings the crescent calls forth with its gentleness, fragility, wistfulness, and pretty loveliness.

Putting aside all these psychocultural ponderings, the bottom line for skywatchers is that the lunar crescent offers some special sights otherwise not visible in the heavens. This is the only phase at which the Moon's light does not necessarily overwhelm our view of a star or a planet that is near it on a particular night (see Sight 22). In this chapter, we will concentrate on two special sights of the lunar crescent: the eerie but lovely glow of earthshine and the golden thread or eyelash of an ultrathin "young moon" or "old moon."

Earthshine

There is a phenomenon that has long been known as "the old Moon in the New Moon's arms." It is mentioned very memorably indeed in the old ballad "Sir Patrick Spens": "Late, late yestre'en I saw the new moon/Wi' the auld moon in her arm,/And I fear, I fear, my dear master,/That we will come to harm." Most of us have noticed this phenomenon at one time or another: it is a crescent Moon with the rest of the Moon's orb much more dimly visible within the crescent's curve. The accepted astronomical name for this glow is evocative but mysterious to the novice astronomer: *earthshine.*

As Leonardo da Vinci seems to have been the first to figure out, earthshine is truly the light of our planet Earth itself seen illuminating the night part of the Moon. The very thinnest lunar crescents don't show this grayish glow, for they are visible only in a rather bright twilight sky, and the glow on the rest of the Moon is too dim to be seen against this background. By contrast, once the crescent becomes thick, its light—reflected sunlight—is usually bright enough to overwhelm the earthshine. The earthshine is dimmer then anyway because the phase of Earth seen from the Moon is always the opposite—or to speak more accurately, the complement—of the phase of the Moon we see from Earth. In other words, when we see a thin lunar crescent, an astronaut looking back from the night side of the Moon would see Earth almost fully lit. Only a crescent-shaped slice of shadowing, exactly the same shape as the lunar crescent we're seeing from Earth, would be visible on the edge of the big, bright Earth seen by the astronaut from the lunar surface. Earthshine is usually easiest to see three to five nights after the New Moon.

A remarkable fact is that earthshine varies in intensity very noticeably depending on whether the illuminated part of the hemisphere of Earth

Venus is bright beside a thin crescent Moon.

facing the Moon is covered with many clouds or few. Clouds reflect far more sunlight into space than do land and sea. If you happen to live on the U.S. East Coast and you see strong earthshine on the lunar crescent at nightfall, it is an indication that there is an unusually large amount of clouds over the rest of the United States and the Pacific Ocean. So simply observing a lunar crescent can enable us to know what wide-scale weather is taking place almost half a world away!

The Quest for the Youngest Moon

The "age" of the Moon is the amount of time that's passed since the moment of the previous New Moon. The New Moon itself is unobservable (except during a solar eclipse, when it is seen in silhouette in front of the Sun). One reason the New Moon can't be seen is that it is overwhelmed by the glare of the nearby blazing Sun. The other reason is that at New Moon it is the dark side—the night side—of the Moon that is facing us, so there is nothing (that is, nothing illuminated) to see. As the Moon starts moving farther out from our line of sight to the Sun, however, two things happen. First, the Moon starts setting long enough after the Sun to be glimpsed very low in the dusk. Second, we start seeing a very slender sliver of the Moon's westward edge being illuminated by the Sun.

This first visible but extremely skinny lunar crescent is called the "young moon." The last visible ultrathin lunar crescent seen as soon as possible before New Moon, in the east as dawn is brightening, is called the "old moon."

All of us no doubt can recall a time when we saw an amazingly slender lunar crescent. But the youngest crescent Moon most people are likely to see just by luck, without foreknowledge and preparation, is maybe about 30 or 35 hours old—that is, 30 or 35 hours past the moment of New Moon. The real thrill comes when you learn what nights of the year there is a good chance to see a younger crescent and then manage to spot a Moon just 24 hours old or even younger. This is the quest for the youngest (or, at dawn, the oldest) Moon.

I first read articles on the youngest moons and the quest to see them in *Sky & Telescope* in the 1970s. These pieces by the chief editor, Joseph Ashbrook, were inspiring. I decided to start looking for ultrathin lunar crescents. My first important observation was a good view of a Moon 24½ hours old. That was very far from any record. But the experience was unlike any other I've ever had, even more than Ashbrook's articles had prepared me for.

I discussed this observation and this Moon in an earlier book of mine. I wrote that the Moon was "thin as a single snippet of pale gold hair falling, falling gently through dusk to the repose of the low forest horizon" and that it was "a slight but dreaming and luminous smile scarcely touching the face of that twilight sky, a face whose tender tones and shades were the only things which could possibly be delicate enough to hold that moon." And yet despite this seeming fragility of the slip of Moon, I wrote that I also had the feeling that this crescent was "an indelible mark, a slenderest sliver of pure celestial beauty that eternity keeps forever from harm or slightest alteration." Last but not least, I felt that my observing site, the most familiar and mundane of places in my neighborhood, was not so ordinary after all if it could hold this wonder in its sky.

Of course, the preceding sentences are largely just a record of my reaction to a young moon. You might be thinking that it was an overly imaginative reaction. What, you might ask, is concretely different about a Moon that is young from the somewhat older ones that most folks have seen at one time or another?

The lunar crescent at such a young age as I saw begins to look like something other than a crescent—something startlingly strange. The thinness is like that of a single golden eyelash. But here's the key: the ends of the crescent begin to get eaten away by the shadows of mountains and highlands on the Moon's edge, so it is not a semicircle anymore—it becomes less and less than 180°. As a matter of fact, these shadows may even cause breaks in the thread of an extremely young (or extremely old) Moon. Such a narrow

A thin crescent Moon.

strand or fiber of Moon is also prone to any turbulence in the long path of air down low, so there may be beadlike pulses of light running along the trembling crescent. This is often better seen with binoculars, but as dramatic as it is, the observer comes to wish for steadier air with which to see a still, perfectly chiseled crescent. I was amazed once to discover that on an unusually steady night I was able to observe in a telescope a number of the severely edge-foreshortened lunar surface features even on a Moon slightly less than one day old. In such a case the illuminated crescent extends only about one-fiftieth the way across the face of the Moon.

The really vital measure of an ultrathin Moon is not age but elongation—that is, the angular separation of the Moon from the Sun, usually expressed in degrees. If the Moon happens to be closer than average to Earth just after New Moon, it reaches a larger elongation from the Sun faster. Even more important is whether the elongation is steep as opposed to shallow. If you live at midnorthern latitudes, the highest young moons are those that occur in late winter to early spring, and the lowest are those in late summer to early fall (the opposite is true for old moons at dawn—they are high around September and low around March). Finally, the timing has to be right for your longitude. In other words, if the New Moon occurs a few hours before sunset, the Moon won't be old enough to see that evening, but the next evening it will be more than 24 hours old.

Do you want some special help in finding very young and very old moons? The best source is the "Young Moon, Old Moon" section of the annual publication *Astronomical Calendar*. On two atlas-sized facing pages, Guy Ottewell plots the position and age of the Moon for several nights before and after every New Moon of the year.

But what is the youngest moon that can possibly be seen—or the smallest elongation from the Sun at which the Moon can be seen? Andre Danjon's study suggested that the smallest elongation at which any of the lunar crescent has not been blotted out by shadows is about 7°. The record for seeing a young moon with a telescope is, at the time of this book's writing, a little more than 12 hours and with binoculars a little more than 13 hours. The naked-eye record is more controversial. A sighting of a Moon just 14½ hours from New Moon was reputed to have been made by two English housemaids on a superbly clear night—of a zeppelin raid—in 1914. But this appears to have been discredited and the accepted modern record seems to be a little more than 15 hours.

Twice I have seen lunar crescents less than 17 hours before New Moon. The first time, the Moon was 16½ hours from New Moon, and I found it in binoculars and then was just barely able to glimpse it with the unaided eye. The other time, the sky was a bit hazier and the Moon required binoculars to see at all. The first of these sub-17-hour Moons, the one I glimpsed with the naked eye, was seen by me from southern New Jersey in superb weather just a little bit later—a little closer to New Moon—than two observers in northern New Jersey that morning. I was a little farther west than they were, so sunrise came to them a bit earlier. I learned that I was very near the theoretically computed limit for seeing a young moon that morning. When I factored in bad weather in other key parts of the country occurring then, I came to an awesome conclusion: that month, I was quite likely the very last person on Earth to see the Moon before it departed from all visibility for a day or two.

Sight 22 Lunar Conjunctions and Occultations

In the loosest sense of the word, a conjunction is any fairly close but temporary pairing or grouping of celestial objects. In its stricter definition, however, a **conjunction** occurs precisely when one celestial body moves to a position in which it has the same celestial longitude or right ascension as a second body. In other words, a conjunction features one object passing due north or due south of another—or, in rare cases, precisely in front of the center of the second body.

As a matter of fact, when a conjunction is so close that one celestial object passes directly in front of another and conceals some or all of the latter, the event is usually called an **occultation**. The related infinitive verb form is "to occult." It literally means "to hide." (When fortune-tellers and other such people speak of "the occult," they mean the supposedly supernatural that is "hidden" from most of us.)

If an occultation is one celestial object hiding another, what is an eclipse? Eclipses of Earth's Moon and the moons of other planets occur when the shadow of another body—usually the planet—falls on the moon, dimming its light. But an "eclipse of the Sun" does involve the body of the Moon hiding part or all the Sun. Why don't we call it an occultation of the Sun? There are two answers. One is that calling the hiding of the Sun an eclipse is a time-honored tradition, dating from several thousand years ago. The other answer is that occultations almost always involve the body that does the hiding being much larger than the one being hidden. But the Moon, as we all know, appears roughly the same size as the Sun in our sky.

A conjunction of the Moon and Saturn.

Lunar Conjunctions

On a number of nights every month, the Moon can be observed near bright planets and stars. If you're using your naked eye, then the planets and stars must be bright for you to perceive them if (1) the Moon is very close to them or (2) the Moon is at a large, bright phase. How close the Moon comes to a particular planet or star varies from month to month. More important, the proximity of the Moon to the planet or star that you actually see depends on where the Moon is in its monthly journey around the heavens at the time your part of the Earth is turned toward it.

The month's agenda of lunar conjunctions can be used by a beginner to identify the stars and the planets the Moon passes. The popular monthly astronomy magazines generally provide diagrams showing the most interesting nights for lunar conjunctions. But the best of all visual guides to what the Moon is near on almost every night of the month is surely the Abrams Planetarium Sky Calendar, a valuable resource for beginners and veteran observers alike (see the sources section in the back of this book). What is that bright point of light left of the Moon tonight? The Sky Calendar will tell you. Which night this month can you (weather permitting) locate Saturn by using the nearby Moon as a guide? The Sky Calendar will tell you that, too (as will columns such as the "Sun, Moon, and Planets," which I write for *Sky & Telescope*).

Lunar conjunctions are a lot more than just useful ways of learning stars and planets, however. It's amazing how much more exciting it is to see two celestial objects near each other than it is to see each by itself. And for night's most spectacular object, the Moon, to be one of the two is particularly good (as long as the Moon is not too bright to completely overwhelm the other object).

Naturally, there are sometimes not just pairings of the Moon with a star or a planet but groupings of the Moon with multiple stars and/or planets. Lines, triangles, and other geometric shapes formed by the Moon and other heavenly lights hold a mighty fascination to us human beings with our built-in propensity for pattern-finding or pattern-making. The possible arrangements of the Moon, the stars, and the planets are endless. I have seen so many in my lifetime as a skywatcher that it would be difficult for me to pick out just a few as outstanding examples—most of them have stirred my interest at least and usually my sense of beauty, too.

Lunar Occultations

Lunar occultations are very different from typical lunar conjunctions. The latter usually involve the Moon a few degrees away from a star or a planet and are best seen with the naked eye's wide field of view or binoculars' fairly wide field of view. In contrast, when the Moon occults a star or a planet, the sight almost always requires a telescope. First, the Moon's glare otherwise overwhelms the tremendously dimmer object that is right up against it. Second and most important, we need magnification to see the details of a star's point of light or a planet's globe poised at the very edge of the Moon. It is especially electrifying to see a star or a planet viewed near a lunar crater or a mountain at the edge.

And then there comes the actual hiding.

When it is a star perched on the Moon's *limb* (another name for the edge of a celestial object), you wait breathlessly at the eyepiece of your telescope—literally. Why? Because you don't want to breathe too hard and risk fogging up the eyepiece even for a second. You don't want to blink too deeply and risk having your eyes shut even a large fraction of a second for fear of missing an event that takes a small fraction of a second. One moment there is an intense speck of twinkling light on the bright, dim (earthshine-lit), or invisible lunar limb. The next moment—poof!—the star winks out of view. Only if the star is a big one—that is, if it is a sun hundreds of millions of miles wide (for instance, a "red giant" such as Antares)—is the fraction of a second slightly, but noticeably, longer.

The best way for me to communicate the drama of these events is perhaps through the following astronomy newspaper column that I wrote *during* an occultation:

LIVE . . . FROM A FIRE ON THE EDGE OF THE MOON

Astronomy column . . . live.

I am writing these words mere minutes after rolling my cannon-like 8½-foot-long telescope across a crust of hard snow to my front yard in a 9 degree F wind chill just after 6 a.m. on Thursday March 3.

The effort and chill were worth it because I was staring through the telescope at the eyepiece-filling, exquisitely detailed ball of the moon and watching something else. I was watching a richest, warmest golden-orange spark of light poise just beyond the edge of that moon, and then look truly like a fire sitting small but perfect on the lunar surface. And then—in thrillingly, almost palpably longer than a moment—that spark seemed to be sucked behind the moon.

What I saw was the first part of an "occultation"—a hiding—of Antares, the bright star which marks the heart of Scorpius the Scorpion. In the 37 years since childhood that I've owned at least one effective telescope, I had seen the rare occultations of four of the five bright planets and three of the four brilliant stars that are capable of being hidden by the moon. Now this hiding of Antares is the fourth of the four bright stars for me.

But I wrote above that I had just witnessed the "first half" of the occultation. Good enough? Yes, but in well under an hour from now—from the minute I'm writing this sentence—Antares will come out from behind the dark part of the moon—in broad daylight. Will I be able to see the star in daylight with my telescope?

I'll find out soon . . .

I've written this column so far quite swiftly. But now I have to go back to polish it. . . .

(A few minutes pass.)

Okay, I've polished, and taken a short break. Now I must go out to watch for Antares to come back. . . .

Success! In the wash of daylight, by the side of the road, as seen in my telescope: a golden speck materialized just beyond the circle of the incompletely lit moon. Antares was back, peeking from 600 light-years away past the rock rampart 1¼ light-seconds away that is our moon. New Jersey will have one more chance, weather permitting, to see an occultation of Antares this year—in May. But this morning was magic.

As things turned out, weather did *not* permit the May occultation of Antares to be observed from New Jersey. Some types of lunar occultations are all the more precious because they are rare.

The Moving Moon

As you watch an occultation, your natural inclination is to think it is the tremendously smaller object—in the case just mentioned, a star—that is doing the moving. The huge Moon in your telescope seems to be the fixed object. Remarkably, the opposite is true.

The cause of the quickly changing, dramatic events called lunar occultations is the motion of the Moon.

Your first reaction is to think of the nightly westward motion of the Moon—its rising, crossing the sky, and setting. But that merely apparent motion is caused by Earth's rotation, and it affects the stars and all celestial bodies. The Moon's true motion in its orbit is *eastward.*

You can prove this to yourself even just with the unaided eye by watching the changes in the Moon's position in relation to a bright star or a planet—either from night to night or even from hour to hour.

Of course, the magnification of a telescope not only enlarges the way an object looks but also effectively speeds up the appearance of motions. So when you watch the Moon approach a star in a telescope, you can really notice the distance between the Moon and the star dwindling. And, once informed, you can appreciate that it is the vast Moon that is catching up to the seemingly tiny star.

Immersions, Emersions, Grazes, and IOTA

I don't think that anything as precise and nail-bitingly dramatic as occultations can ever be regarded in a lackadaisical manner. They are suspenseful, sudden, and gripping. Nevertheless, we can say that observing occultations can be carried out just for fun and with a minimum of preparation of attention to detail. Or, alternatively, an amateur astronomer can perform some serious scientific research by accurately timing certain occultations.

Dedicated occultation watchers need to know the predicted times of immersion and emersion. *Immersion* is the disappearance of the object behind the Moon. *Emersion* is the reappearance—the coming back out—of the object from behind the Moon. The opportunity for science and the need for accurate timing happens when there are several stages to immersion and/or emersion. In one case, the several stages represent two members of a close double-star system being hidden or revealed. No telescope in the world may provide a sharp enough ultrahigh-magnification image to show the single point of light as two stars. But an amateur astronomer with a small telescope may glimpse the light of the star first fade and then, very soon after, disappear completely—evidence that the point of light is really a double star. Even in today's high-tech world this is the way that the doubleness of some stars is discovered.

The other case in which there can be more than one moment of immersion and emersion is even more dramatic: it is a *grazing occultation.* The observer is seeing the star (or planet) pass tangentially along the edge of the Moon. The star can wink on and off a few times or even dozens of times as it is alternately hidden behind lunar mountains and highlands and revealed through lunar valleys and lowlands. Even if the variation in topography seems almost imperceptibly slight to direct observation, it will be enough to cause a star to fade and disappear in a grazing occultation. Over the years, occultation watchers have actually gained much precise information about the topography in the limb regions that lunar geologists weren't able to obtain in any other manner.

Well, actually, I should also say occultation timers and videotapers, not just occultation observers. The premier organization of amateur astronomers that performs this work is IOTA—the International Occultation Timing Association (see the sources section in the back of this book). Dr. David Dunham, its guiding light for decades, and its other most devoted members live a remarkable existence, traveling to remote locations for grazes up to dozens of times a year. By remote I not only mean a great distance from home but also sometimes to off-the-beaten-track locations, far from the main roads. The reason they travel so far is that the zone of territory in which a grazing occultation can be seen may be long, but it is only a few miles wide—the amount of variation in elevation that exists on the lunar limb along which the graze is occurring.

By the way, another scientifically important kind of occultation that the IOTA watches and times are those of asteroids by stars. Exactly where these will be seen on Earth is hard to predict, but if observers are positioned across a large geographical area, some of them should catch sight of a star either disappearing or suddenly being replaced by a possibly dimmer object that has just moved in front of it—an asteroid. If observers at several different locations record when and how long they see the dimming, it may be possible to determine the diameter and shape of a previously little-known asteroid. Amateur observers of these occultations have also discovered moons of asteroids when the star winked out of view briefly before or after the asteroid itself was supposed to hide it.

Lunar Occultations of Planets

As dramatic as the Moon's occultations of stars are, they are not as individually unique or as fascinating in their strangeness and richness as the other major kind of lunar occultations: those of planets.

Each planet presents different opportunities for wonder to occultation observers.

Mercury glows little in a twilight sky, sometimes with colored clouds in view, at the edge of a vast skinny hook of Moon.

Venus is so bright and intense compared to the surface brightness of the Moon that it can be seen with the naked eye in the act of rapidly fading as its form spends seconds passing behind the lunar limb. Once, on a crystal-clear morning the day after Christmas, I watched in a telescope the horns of the Venus crescent suddenly start materializing in blue sky as if from nowhere because they appeared from behind the dark—and against that bright sky, invisible—side of the Moon. Then I quickly pulled my head back to see with my naked eye the wondrous sight of the Venus point of light rapidly brightening as more and more of the planet's globe emerged from behind the Moon.

I have seen Mars float ever so close to the edge of the Moon before and after sunrise when the planet was unusually close to Earth. Not only did its globe's predominant orange contrast magnificently with the yellow lunar landscape. Its dark surface features remained sharply visible and its polar cap became even more shockingly white when seen next to the blue sky.

Jupiter's big, banded globe forms an outrageous bump on the side of the Moon as it begins an immersion that takes more than a minute. And this event is accompanied by immersions and emersions of Jupiter's four big bright moons.

The several occultations of Saturn I've seen were as awesome as you might imagine. One presented, in the late morning sunlight between rapid blowbys of thin cloud, a sudden ghostly sight of a seeming crescent Saturn-globe-and-rings steadily thinning into nothing. The view reminded me of paintings of Saturn seen in the Saturnian moon Titan's imagined blue and white cloud-streaked sky by the pioneer space artist Chesley Bonestell in the middle part of the twentieth century. Bonestell's picture and my sight were ones of awesome dreaminess and tranquility. A few years earlier, I also observed—again through clouds and breaks in clouds—a nighttime grazing occultation of Saturn by the Moon. At one shocking moment, I realized I was seeing the tiny bump of a lunar mountain right in front of one part of the rings. I was on a ladder to view this event through a tall telescope of mine. When it was over, I dropped off the ladder in such a daze of delight and wonder I hardly knew which world I was in.

PLANETARY CONJUNCTIONS

Sight 23

If you're reading these chapters in order, you probably think that the previous one expressed about as much excitement and wonderment about conjunctions and occultations as possible. I was sharing visions of mine like that of Saturn wading, rings-deep, along the bumpy edge of a huge golden Moon.

But by now in this book you should have learned that each of the best sights in astronomy has its own special attraction, power, charm, or thrill. This is true even of ones that might seem similar. For instance, wouldn't conjunctions of one planet with another be similar to the sight of the Moon near a planet? Yes and no. The Moon is its own big, brutally bright traveling show, whipping all the way around the heavens every four weeks. It can seem too common and too overbearing (it is so bright it actually throws into obscurity almost all celestial objects it gets near). But the planets—ah, now there are objects with a relative slowness of motion that lends them an added nobility and grandeur—especially when we ponder the years, decades, and sometimes even centuries that pass before the clockwork of two or more planets' motions bring them together at a certain time and place in the heavens.

In this chapter, we will consider not just pairings of one planet with another but also larger arrangements of greater numbers of planets. We will also include conjunctions that involve a planet passing near a bright star or a star cluster. The basic concept of what is happening in these events is simple: a planet is near to, and forming a pattern with, one or more other celestial objects (excluding the Moon, which we considered in the previous chapter).

Bright Conjunctions and Close Conjunctions

From the standpoint of sheer light output, the most impressive planet-planet pairings are those of Venus and Jupiter. Even when these two are in just the same general area of sky together they draw the attention of the public. When Venus and Jupiter get within as little as 5° of each other in plain sight, the police and airports will often start getting reports of UFOs from people who don't understand what they are seeing. Ironically, an ultraclose conjunction of Venus and Jupiter may have been the central in a series of conjunctions that was the Star of Bethlehem in 3–2 B.C. That conjunction was so

The conjunction of Venus, Jupiter, and Mars on June 13, 1991.

close that the naked eye saw the two masses of light merge into one. I've seen for myself what a prodigious sight it is when these two mighty planets come within a fraction of a degree of each other. You might be lucky enough to get a good view of Venus and Jupiter within a degree of each other only a few times in your life. Don't miss this conjunction!

Of course, both objects, or even one object, in a conjunction does not have to be bright for the event to be thrilling to see. After all, the essential quality of a conjunction is closeness, so two planets being extremely close together in our sky counts for a lot. Even in discussing the brilliant Venus-Jupiter conjunctions, I gave special emphasis to the closest of them.

Many of the conjunctions I remember most vividly are very close ones. In June 2005, I viewed a pairing of Venus and a much less brilliant planet, Mercury. Conjunctions of these two inner planets are common but often too deep in the solar glare to see well. On June 26, 2005, however, I was able to see Mercury plainly with the unaided eye, even though it was rather low and twilight was still rather bright. And I was absolutely delighted to see it because that evening Mercury was only about 0.17° from Venus. As traffic coming back from the Jersey shore roared by on the highway next to the deserted churchyard where I stood, I felt as though I was in this world but wonderfully not of it; I was transported out of myself by the sight of two worlds seemingly almost touching, one precise planetary point of light tucked right under the much brighter one.

By the way, despite the fast motions of Venus and Mercury, every few years they can match their movements closely enough to keep them within a degree or two of each other for as long as a few weeks. That is more than just a conjunction, it is a conjunction dance, and it is lovely to watch.

What has been my favorite conjunction between two planets? A strong candidate is the very close one of Mars and Saturn back on June 4, 1978. The two were just past their closest together—about 0.1° apart—as evening fell here in the eastern United States. The night was very clear and the two were far from the Sun, so it was possible to enjoy them high in a dark sky for hours. When two stars or planets of different colors get close to each other—actually even within a few degrees may suffice—the hues become beautifully exaggerated by their contrast with each other (this perceptual effect is well known to double-star observers—see Sight 45). That night in 1978 the gold of Saturn and the orange of Mars were powerfully enhanced by each other. Direct inspection with the naked eye showed that there was still a gap between them. But our vision is very sharp only within a few degrees of its center. If my eye (or anyone else's that night) looked even slightly away to take in more of the overall scene, the reduced sharpness of vision made the two planets look as if their rays *were* actually touching each other.

The view in my telescope of this Mars-Saturn conjunction was also amazing. Most astronomical telescopes are capable of providing a field of view that is more than a degree wide—so when a conjunction is at least that close you get to see the two planets in the eyepiece together. The problem is that wide telescopic fields are generally achieved with fairly low magnification—perhaps not enough to show the globes of the planets as more than tiny dots or specks. That night of the ultraclose Mars-Saturn conjunction, the two worlds were so close together I was able to fit them both into one field of view in my 8-inch telescope at a power of 200× (rather high magnification). In a single view, I could behold Saturn's globe, rings, and brighter moons together with a Mars on which a polar ice cap and dark markings were clearly visible—an absolutely unique sight in my life.

Occultations by Planets

What would be the closest conjunction of planets possible? That would be when one looks in the telescope and sees the disk of one planet pass partly or completely in front of the disk of another. Such an event is called a "mutual occultation" of planets. Only a few of these were known up until the late 1970s, but then my friend Steve Albers was able to calculate more than a dozen between 1557 and 2230. Unfortunately, no mutual occultation of planets occurs between 1818 and 2065—and even the 2065 one will be extremely difficult to observe.

A little more common are planetary occultations of fairly bright stars. In the 1970s, Jupiter occulted a 2nd-magnitude star (Beta Scorpii) and Mars occulted a 3rd-magnitude star (Epsilon Geminorum). The latter was a wondrous event that was well placed for observation by myself and other amateur astronomers in the eastern United States on April 8, 1976. In the late 1980s, a dim naked-eye star passed right behind the rings of Saturn, a marvelous sight in good amateur telescopes.

How often does a planet occult a 1st-magnitude star? Apparently, only once every few centuries. Such an event—Venus occulting Regulus—did happen and was observed back in 1959. But it's important to remember that *extreme* closeness is not necessary to make a conjunction beautiful and memorable. There are many lovely pairings of Venus and that heart-star of Leo, Regulus. A visible one takes place almost every year. Furthermore, because Venus comes back to almost the same position in the heavens every eight years (give or take a day or two), a near-identical replay of any particular Venus conjunction with a star occurs every eight years. In one of these, Venus is close to Regulus just before a dawn in early September. It was at one such occurrence of this conjunction that I found myself watching Venus and Regulus through a small hole in an otherwise completely overcast sky. Even to the naked eye the planet shined lustrously yellow and the star vividly blue-white—because of their proximity to each other their apparent hues are exaggerated. I watched them with naked eye, then with binoculars. And then, at that silent hour before dawn, I heard an astonishingly unexpected sound: raindrops beginning to patter on leaves. The closely paired planet and star burned on brightly while a rain shower strengthened a bit and sounded all around me.

Another Venus thrill to look forward to every eight years is the close passage of Venus by the Pleiades. There are several times in the eight-year cycle of recurring Venus appearances that it goes somewhat near that loveliest of clusters. But the really close encounter is the one that happens in late March/early April once every eight years. In 1996, Venus skimmed less than one cluster–diameter south of the Pleiades (with the bonus of both a bright Comet Hyakutake and a total lunar eclipse on one of the big nights of the passage!). In 2004, Venus skirted even closer to the south of the cluster.

And there is great news for relatively young readers of this book: Every eight years Venus will travel farther north and finally, in 2026, it will start passing right among the main stars of the cluster. Not until 2060 will Venus actually pass entirely north of the brightest stars of the Pleiades.

Another bright star cluster—M44, the Beehive cluster (see Sight 27)—is less spectacular than the Pleiades, but it is much closer to the central line of the zodiac. That is why the Beehive is so frequently crossed by planets, passages that are often quite lovely in binoculars and telescopes.

Trios, Conclaves, and Triple Conjunctions

Each combination of each planet with another, with a bright star, or with a star cluster is different and interesting in its own right—though, as we've just seen, especially interesting if there is a lot of brightness or closeness involved. But the possibilities don't end there. What about gatherings of more than two planets or of one or two planets with one or more stars? What about having not just one conjunction of two planets but a series of conjunctions of the two?

A temporary gathering of three celestial objects within a circle less than 5° wide is called a *trio.* This term, invented by the Belgian astronomical calculator Jean Meeus, has gained wide acceptance. But getting four bright celestial objects—especially four bright planets—into a relatively small region of sky is so rare that this kind of gathering doesn't even have a name of its own. When it does happen—or when all five of the bright planets are visible in the same overall direction in the sky—various writers have referred to the event as a "conclave of planets" or "parliament of planets." The most famous of these in our time occurred in the early 1980s, when for the last time in several centuries not just the five bright planets but all eight of Earth's fellow planets, including Pluto, could be observed above the horizon at the same time. Fortunately, the still rather exciting event of getting all five so-called "naked-eye planets" visible above the horizon at once actually happens every few years—it's just that we usually have to see a few of the planets low in the west and a few low in the east, either at dusk or dawn.

Quite different from a trio is a triple conjunction. A triple conjunction is a series of three conjunctions between two objects occurring in a matter of just months. It's not that rare of an event if one (or both) of the two swift-moving inferior planets (planets closer to the Sun in space than Earth is) are involved. Over the course of just a few months, these two planets can shuttle out to their maximum angular separation from the Sun in dusk and then back to the Sun and out to their maximum angular separation from the Sun at dawn. In doing so, they can fairly easily pass each other or another planet three times in less than a year. What is tremendously more rare is a triple conjunction of the superior planets (planets farther from the Sun in space than Earth is).

The most famous triple conjunction of superior planets is that of the two giants of the solar system: Jupiter and Saturn.

Most Jupiter-Saturn conjunctions are single ones. The faster planet, Jupiter, takes almost twelve years to circle the heavens but then needs an additional eight years to make up the ground that Saturn has been covering. Thus, Jupiter-Saturn encounters occur at twenty-year intervals (the next is not until 2020). At least when Jupiter does encounter Saturn the two are so slow that they remain relatively close to each other for months.

But usually every seventh encounter of Jupiter and Saturn offers us a greater thrill: a triple conjunction. Why does this sometimes happen? Each year when Earth begins to catch up to a superior planet there is a period of weeks or months when the planet seems to drift backward (westward) relative to the background of stars (the true orbital motion of the planets, remember, is east relative to the starry background). This backward or *retrograde motion* is merely an effect of perspective. It is comparable to something we all experience when driving on a multilane highway. As we pass a slower-moving car, that car appears to drift backward relative to distant background scenery, even though we are both moving forward. The same thing occurs when Earth passes a slower planet, with the distant stars serving as the background scenery. In any case, if one planet passes another at the right time in this Earth-caused yearly sequence of direct (eastward) then retrograde (westward) then direct (eastward) motion, the two planets can appear to shuttle back and forth through a series of three conjunctions as seen from Earth.

The triple conjunction of Jupiter and Saturn usually occurs only after about 140 years. One of these events occurred in 7 B.C. By at least the early seventeenth century (maybe much earlier), this conjunction was being suggested as an explanation for the identity of the Star of Bethlehem (which the Bible implies was seen on at least two different occasions by the Magi).

The last Triple Conjunction of Jupiter and Saturn took place in 1981, so the next is far in the future. But part of the grandeur of conjunctions is watching the majestic progress of planets toward their next meeting—for instance, Jupiter first reaching the opposite side of the heavens from Saturn and then beginning the slow approach.

The Greatest Triple Conjunction and Trio

An equally rare, but to my mind even more visually impressive triple conjunction, is that of Mars and Jupiter. This has received little attention. Mars normally catches back up to Jupiter in our sky and has another conjunction with the latter after a little more than two years. These conjunctions always feature Mars relatively dim and small because it is never more than about 90° from the Sun. Did I say "always"? I meant every time except for once every 143 years. In that exceptional case, the planets have a triple conjunction, with the middle event of the series occurring with both Mars and Jupiter near opposition—opposite the Sun in the sky and therefore visible all night long at their brightest and biggest.

This central event in a triple conjunction of Mars and Jupiter last happened on the night of February 29/March 1, 1980. The planets were just a few days past both coming to opposition within 12 hours of each other. And the conjunction of them that I observed on that Leap Day evening and the rest of the night was one of the most monumental I've ever seen. That night I was outside Binghamton, New York, and the temperature dived to –23° F. Up in the sky the two brilliant planets stood right beside the precisely Full Moon—and together proved the only night-sky sight I have ever seen that actually exceeded even the sheer power and impact of the Full Moon.

What could surpass this central conjunction of the Mars-Jupiter series and the way it had upstaged the Moon? The third and final conjunction could not be as bright but it was, incredibly, even more beautiful. Far more beautiful—possibly the most beautiful gathering of celestial objects I've ever witnessed. The reason was that this time, even though Mars was a lot dimmer than before, the two planets formed a tight trio with 1st-magnitude star Regulus, which marks the heart of Leo the Lion. The big evening—that of May 3, 1980—was wonderfully clear. The three lights were high in the south at nightfall. All evening long, they clustered within a circle little more than 1.5° apart. Jupiter shined at magnitude –2.2, Mars at +0.2, and Regulus at +1.3. To the eye or eye-brain system, this grouping was so dense with bright objects that it became far greater than just the sum of its parts. This was especially true because of the extreme enhancement of their different apparent colors that resulted from their proximity to each other. After full darkness had fallen, my unaided eyes saw the heart of Leo the Lion overflowing with the radiance of a magnificent triple "star" of vivid gold, orange, and blue-white.

Sight 24 Bright Comet Close-Up

In Sight 13, we looked at the overall spectacle of a bright, long-tailed comet. Our field of view was approximately 15°-to-50° wide (though some comet tails are even longer than that!). Our optical instruments of choice were the naked eye and, to a lesser extent, binoculars.

There is a lot more to comets than their famous trademarks: their tails. The tail of gas and dust is emitted from the solid part of a comet, the tiny icy "nucleus" being heated by the Sun. But we can never directly see that nucleus from Earth: not only because it is too small to appear as a sizable object (larger than a point of light) but also because it is always shrouded in the shining cloud that with it forms the *head* of the comet. This cloud is known as the *coma* (Latin for "hair" and used here in reference to the fuzzy appearance of this cloud of light). The coma can be spectacularly visible, even as bright as the brightest stars. But to get a detailed view of the structure of the coma and of the brightest part of the tail, that part near the head, we need to use substantial optical aid—a telescope.

Colors in Comets

When a comet moves in closer toward the Sun, the icy cometary nucleus is heated. In the vacuum of space, the ice—mostly water ice but also significant

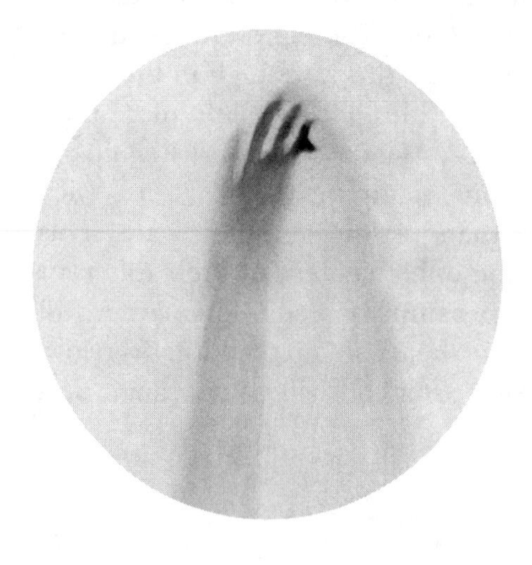

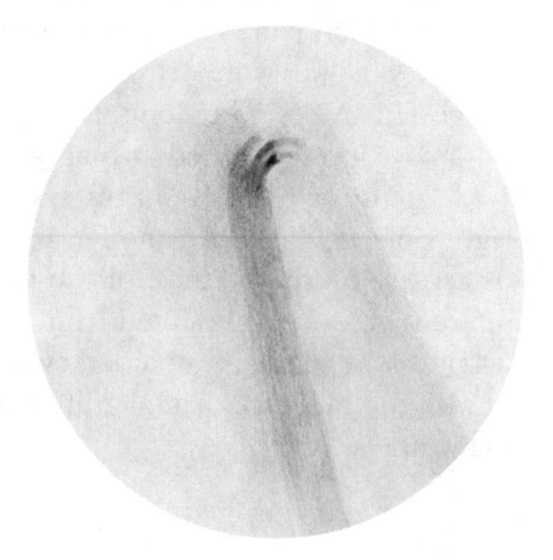

Close-up views of the hoods in Hale-Bopp's coma on March 9, 1997 (left), and March 12, 1997.

amounts of frozen carbon dioxide and carbon monoxide—goes directly from solid to gas when heated (this process is called "sublimation"). Some of the gas just shines by reflected sunlight, but much of it is fluoresced by solar radiation and begins to shine on its own with a bluish tint. This blue tint is often observed faintly in telescopes when a coma begins to develop (an 8-inch or 10-inch telescope may suffice to start showing this color in a comet that has brightened to 6th or 7th magnitude). But the escaping gas may carry with it a lot of dust. The dust will shine by reflected sunlight; therefore, in a dusty and bright enough comet, it will take on a yellowish hue. In a few of the brightest comets (especially Comet West), I have seen the two colors—blue and yellow—combine in places to shine with a green hue. In some very bright comets, the dust may even take on a slightly reddish cast. I saw blue contrasting with such red particularly well in the head and near-tail region of Comet Hale-Bopp, and in Comet Hyakutake when it was nearing the Sun. (A few nights before Hyakutake's closest approach to Earth, just before its inner coma swelled out impressively and produced a super "jet," I observed this inner coma or "central condensation" with a 10-inch telescope and noted it as having a distinctly orange hue.)

Usually, only photographs can reveal the blue of a comet's *gas tail* (also called *ion tail*) or the yellow or red of its *dust tail.* But in the brightest parts of the two kinds of tail, nearest to the coma, the color may be directly visible in telescopes if the comet is very bright and the tails are well separated.

Structures in the Coma and Near-Tail

Even a faint coma without visible color will usually display a brighter area at its center. The extent to which a coma is concentrated toward a center is called the *degree of condensation* (or DC) and ranges from 0 (coma completely diffuse, no condensation toward the center) to 9 (a starlike coma). The *central condensation* in the coma may in some comets become a sizable and rather hard-edged disk (in some comets—such as Comet West—truly planetlike in appearance!). Or it may contain a starlike point of light. In either case, disk or point, the object can be called the *apparent nucleus.* It is always much larger than the true nucleus (which is only a few miles across). The apparent nucleus must be the densest cloud of material nearest to the nucleus.

Even a comet too dim to see with the naked eye may show a faint *sunward fan* of light in a telescope. Fairly high magnification may reveal one or more little tuftlike features called *jets* coming out from the central condensation. These are fountains of gas and/or dust shooting up from the nucleus when compact active regions of particularly volatile ice—perhaps a part of the surface that is naked of dust—are heated enough by the Sun. Expanding outward from the inner coma in circular shells may be *halos*—also called *envelopes* or *hoods*—of gas and/or dust. (For many weeks, the great Comet Hale-Bopp displayed several layers of hoods that were easily visible in amateur telescopes—see the figure on page 134.) Halos may result from linear or ring-shaped active regions on the nucleus, or if they are seen edge-on may mimic jets in appearance.

A large comet that gets much closer to the Sun than Earth will have the ionized part of its gas condensed and driven back by the solar wind, which consists of charged atomic particles from the Sun. When this happens, we see the inner coma take on a dramatic parabolic shape. Between the two arms of the parabola, extending back from the nuclear region there may appear either a *shadow of the nucleus* (a narrow strip of reduced brightness) or a *spine* (a very thin line of increased brightness). As you might guess, the so-called shadow of the nucleus is not really a shadow of the tiny icy nucleus but just an area that receives less sunlight because it is directly behind (anti-sunward of) a particularly dense central condensation.

THE HYADES STAR CLUSTER AND ALDEBARAN

Sight 25

There is only one star cluster in the heavens that looks larger than the Hyades. It is the Ursa Major Cluster, whose stars are scattered over too large an area, and it is interspersed with too many other (mostly background) stars to properly appreciate. The Hyades cluster has a large apparent size. But it is still a remarkable coincidence that one of the stars that happen to lie in front of it is a 1st-magnitude one. Aldebaran is that star, 65 light-years from Earth, while the Hyades cluster is centered about 150 light-years away.

The Hyades star cluster and Aldebaran have already been discussed in Sights 9, 15, and 17. In those chapters, they were considered as part of a marvelous group of constellations (the Orion group), as part of one of the brightest and most interesting individual constellations (Taurus), and as one of the brightest stars and the naked-eye cluster it shines in front of.

Here, we'll take a quick look at some of the close-up wonders of this star and cluster—for several of the interesting qualities of them do require the greater magnification and light-gathering power of binoculars or a telescope to appreciate.

An aurora in Iceland.

The transit of Venus in 2004.

The Pleiades.

The Great Nebula in Orion.

Steve Albers produced this image of the total solar eclipse of July 11, 1991, by using a computer program of his own creation to digitally combine five different photographs taken by Dennis di Cicco and Gary Emerson.

Comet West in March 1976.

Streamers in 1996's Comet Hyakutake.

Comet Hale-Bopp on March 31, 1997.

Orion with Leonid meteors at a strong Leonid outburst in 2001.

The double star Albireo in Cygnus.

The Great Galaxy in Andromeda.

The Hyades (lower left) and Pleiades star (center) clusters.

Fill Those Wide-Field Binoculars

How big is the V- or arrowhead-shape outlined by the brightest Hyades and Aldebaran? At least 5° across—so you'll need wide-field binoculars to contain the whole pattern. And what a glorious sight when you do! Aldebaran's orange—some take poetic license and say rosy—hue becomes more apparent. So, too, do the oranges and yellows of some of the Hyades (unlike the Pleiades, this cluster is old enough for a number of its members to have already become red giants).

There's an amazing number of bright objects in your binoculars when you look at Aldebaran and the Hyades. Aldebaran is the fourteenth brightest star in all the heavens, at magnitude 0.87. The combined brightness of the Hyades is even greater. But, of course, it is the parceling out of that light into so many different naked-eye stars that makes the Hyades so impressive. Consider the following statistics: Five of the Hyades shine between magnitude 3.4 and 3.8. Six more are magnitude 4.5 or brighter. Five more are between 4.5 and 5. That's sixteen stars that should be bright enough to detect with the naked eye even from many small cities and moderately light-polluted suburbs.

Ten more Hyads are brighter than the traditional naked-eye limit in dark skies, magnitude 6.5. There are more than about 130 Hyades brighter than 9th magnitude and thus within the range of good binoculars.

If you have a good, dark rural sky, or a pair of binoculars in less good skies, you will see that there are a number of (mostly fainter) Hyads that add to the classic V of the main pattern. I see an N or (more often) a Z, formed by the V plus an extra stroke of stars stretching south-southwest from Aldebaran or farther from the wide double-star Sigma Tauri (roughly southeast of Aldebaran).

Double Stars of the Hyades

Sigma Tauri is just one of the fine stellar duos you can find in the Hyades. Its two blue-white stars shine at 4.7 and 5.0 and are separated by about 7'. The brightest Hyad (true member of the cluster) is Theta-1 Tauri, which shines at 3.4, just over 5.5' from magnitude 3.8 Theta-2. The two are yellow and orange. Can you split these pairs with your unaided eye? If you can't, then try the wider pair formed by Delta Tauri (magnitude 3.8) and 64 Tauri (magnitude 4.7).

The really bright Hyad that is farthest from the main flock (the V) in the sky is Kappa Tauri. This magnitude 4.4 star is several degrees north of the V. It has the magnitude 5.4 star 67 Tauri 5.5' from it.

There are at least ten other Hyades doubles that can be split with amateur telescopes of various sizes.

More Aldebaran

While you admire Aldebaran in your binoculars or telescope, you can ponder some amazing facts about it that we haven't yet discussed.

First, consider that Aldebaran is now about 350 times as luminous as our Sun and about 40 times wider—but that our Sun should someday (a billion years in the future?) become a star much like Aldebaran. (By the way, while the Sun takes a little less than a month to rotate, Aldebaran takes almost two years to do so.)

Of all 1st-magnitude stars, Aldebaran is the one receding from us most rapidly. In less than 500,000 years from now, this 1st-magnitude star will have pulled far enough away to have dimmed to 3rd magnitude. In the slightly more recent past, however, Aldebaran played an even more illustrious role in Earth's sky than it does now. From about 420,000 to 210,000 years ago, a much closer Aldebaran reigned as the brightest star in our sky. At closest approach, the star was only about 21.5 light-years away and burned with about the same apparent brightness that Sirius now does.

THE PLEIADES

Sight 26

If there is a lovelier naked-eye sight in astronomy than the Pleiades, or a more distinctive naked-eye vision of stars, I don't know what it is. The astonishing naked-eye view of six, seven, or more moderately bright stars gathered together just about as closely as they could be without losing their individual recognizability—there is no other sight like this in all the heavens.

We've already noted the Pleiades in the wider context of their presence in the Orion group of constellations (Sight 9) and Taurus the Bull (Sight 17). Now, let's take a closer look. Great new wonders of the cluster are revealed by binoculars and telescope. But first, we turn to the magnificent naked-eye view of them we just alluded to. This time, we fix the gaze of our unaided eyes not on the Pleaides in their overall setting but on the cluster itself. (For a stunning color image, see the photo insert.)

The Lore and the Naked-Eye Beauty

Just how powerful the naked-eye sight of the Pleiades is can be shown by the role this star cluster has played in legend, poetry, and even timekeeping throughout history. The Pleiades appear in great literature and poetry all the way from the Bible and Homer to Alfred Lord Tennyson's "Locksley Hall," and to more recent verse. "Canst thou loose the bands of Orion, or bind Pleiades with sweet influences?" asks God of his critic Job. Thousands of years later, Subaru automobiles display the company's emblem, which is a stylized image of the Pleiades (in Japanese, the word *Subaru* means "Pleiades"). Perhaps almost 2,000 years ago, there flourished a culture that based its very calendar on the rising and setting times of the Pleiades. I'm talking about the mysterious Druids, whose year-beginning holiday, Samhain, eventually became Halloween. Samhain was held each year when the Pleiades began rising at nightfall.

Why such interest in, even veneration for, the Pleiades throughout history? A look of your own tonight will show you why. Even after all these years as a skywatcher, I am still sometimes startled when I step outside and glimpse, out of the corner of my eye, a certain eerie sight that has just emerged over the tree line or out from behind some clouds. What in the world—or, rather, out of this world—is it? A piece of sky larger than the Moon (though it seems smaller and far tighter) alive across its entirety with multiply twinkling lights.

The Pleiades.

A lovely bunching of stars, a rich handful of stellar gems, a pocketful of radiance: the Pleiades star cluster.

Everyone is first awed by the strangeness and splendor, then moved to affection by the gentle loveliness of the Pleiades. Almost every culture in the world and throughout history has imagined the cluster as something gentle and delicate—a group of maidens, a flock of doves, or a mother hen and its chicks. The cluster is best known in its ancient Greek imagining: the Seven Sisters. Look with the naked eye and you will see that these sisters—the stars of the Pleiades—twinkle almost but not quite together, a beautiful effect. Sometimes you can actually see by their delicate trembling the passage of small waves or cells of turbulence through the atmosphere. Add to this the traces of strands of glowing gas around these stars that photos plainly show, that telescopes can subtly display, and that the naked eye sometimes just might possibly glimpse. What you have then is what Tennyson noted in his famous lines from "Locksley Hall": "Many a night I saw the Pleiads, rising thro' the mellow shade,/Glitter like a swarm of fireflies tangled in a silver braid."

The shape formed by the main stars of the Pleiades is something like that of a teacup (an appropriately delicate and pretty object). But the form of the cluster is most often said to look like a tiny dipper of stars—and indeed more than a few novices suppose that it must be the Little Dipper. But the Little Dipper is actually very much more spread out and less conspicuous than the Pleiades. Even the Hyades cluster, though possessing more bright stars than the Pleiades, is spread over a large enough area to reduce the impact of its sight somewhat. The Pleiades possess just about the ideal degree of concentration to produce a maximum effect of richness and splendor to the naked-eye observer.

Counting Pleiads and the "Lost Pleiad"

How many Pleiads can a person see with the naked eye? The answer depends partly, of course, on how bright these stars are.

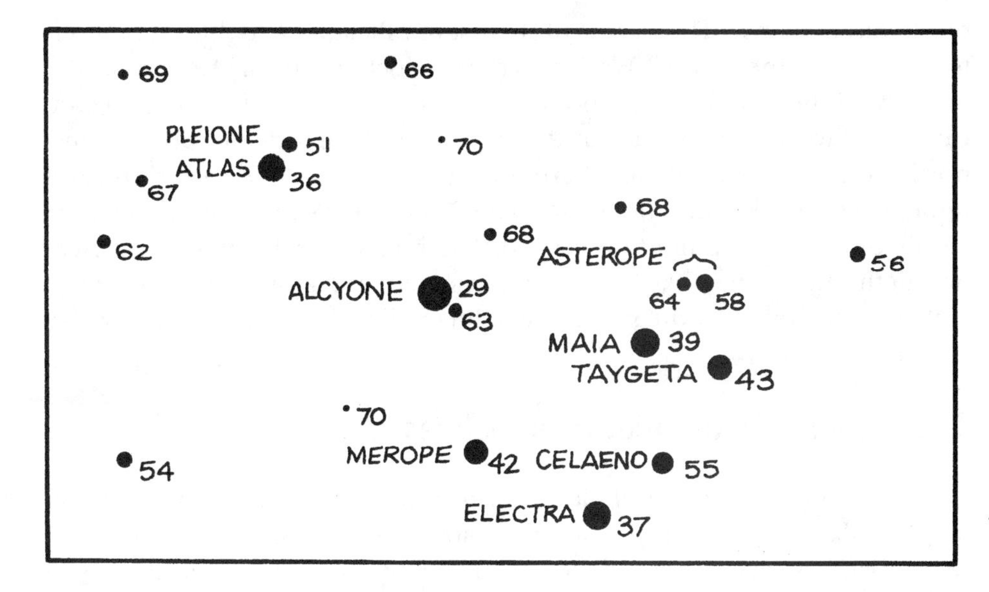

The stars of the Pleiades.

The brightest, which outshines any of the Hyades stars, is magnitude 2.8 Alcyone. It is in legend the mother hen around which the other Pleiads crowd like chicks. It is the star that marks the juncture between the handle and the bowl of the tiny dipper of the Pleiades. Including Alcyone, there are nine Pleiads brighter than magnitude 5.6—hence technically bright enough to see with the unaided eye at a fairly dark country location. Yet most people see either seven Pleiads (hence the name "Seven Sisters")—or six. This discrepancy of one star has led to a beautiful worldwide legend of the "Lost Pleiad." The nineteenth-century poet Alfred Austin alluded to it in these wonderful lines: "The Sister Stars that once were seven/Mourn for their missing mate in heaven." Why do some people see seven Pleiads and others only six? Not only observing conditions but also sharpness of vision plays a role, for several of the Pleiads are quite close to one another. Note especially the two stars that mark the end of the handle of the Pleiades dipper. The brighter is magnitude 3.6 Atlas and the dimmer is magnitude 5.1 Pleione. The two are located about 5' apart, so they are a difficult split for subpar vision or sky conditions. Pleione is known to be at least slightly variable in brightness, and some writers have speculated that it used to be a lot brighter so that everyone easily saw seven stars. But we probably don't have to invoke that special circumstance to explain why one of the seven seemed to be missing to some people at some times.

By the way, there are nine Pleiades stars with proper names. Atlas and Pleione are actually the names of the parents of the Seven Sisters in Greek mythology (Atlas is the famous world-shouldering giant who gives his name to our books of maps). The names of their seven Pleiades daughters are given to the stars of the "bowl." Working counterclockwise from Alcyone at the juncture of the "handle" and "bowl" the stars are Alcyone, Merope, Electra, Celaeno, Taygeta, and Maia with Asterope (actually a wide double star) just north of Taygeta and Maia. There are other, dimmer Pleiades stars, and under excellent sky conditions, sharp-eyed observers can see ten, twelve, or even more of them without optical aid. The record may belong to the great twentieth-century deep-sky observer and amateur astronomer par excellence Walter Scott Houston, who glimpsed eighteen Pleiads with his naked eye.

The Pleiades in Binoculars and Telescope

Anyone can glimpse dozens of Pleiades stars with binoculars. And a small telescope in dark skies will reveal a few hundred stars here. It's important to remember, however, that a wide telescopic field is needed to fit in the whole cluster. Even just the nine brightest Pleiads are spread over more than 1° of sky.

The most impressive views of the Pleiades are those in large binoculars and rich-field telescopes. The cluster still seems gloriously crowded together in richness, but close pairs the naked eye had trouble separating are now easily split. You can detect the tiny triangle of stars very near Alcyone and the brighter curve or cascade of stars running south then southeast from Alcyone. The apparent brilliance of the stars is increased by the light-gathering power of binoculars and telescopes, and the blue tinge of these hot, young, luminous stars becomes noticeable.

The Pleiades are located about 400 light-years from us, more than 2½ times as far as the Hyades. Study of the spectra of the Pleiads shows that they are rapidly rotating stars. This and other signs of youth led scientists to think that the traces of nebulosity around these stars were remnants of the cloud in which they were born—sort of the swaddling clothes of the infants they recently were. Astronomers now believe, however, that this nebulosity is simply a cloud that the cluster is passing through and temporarily lighting. Whatever its origin, the Pleiades' nebulosity is elusive for beginners. You certainly need a dark, clear sky and clean optics in your telescope. If you then think you are seeing patches of dim glow near a few of the Pleiads, try looking at the Hyades. If the Hyades stars also seem to have this glow around them, then you are just seeing condensation on your optics being lit by the starlight. When you think you see nebulosity associated with the Pleiades, where do you see it strongest? If it seems strongest in a patch extending south from Merope, the star in the bottom of the Pleiades bowl that is closer to the handle, then you are probably detecting the nebulosity.

Other Very Bright Large Open-Star Clusters

Sight 27

In the past two chapters, we visited the two brightest, most impressive naked-eye star clusters. We saw that both the Hyades and the Pleiades were best observed with the naked eye and with an optical instrument (binoculars or a rich-field telescope) that provided a very wide field of view.

The same conditions for optimum visibility apply for several other very large open clusters. These clusters are not as bright as the Hyades or the Pleiades and not as impressive. But each has its own distinctive aspects that are interesting and any one of these clusters can still be considered one of the best sights in astronomy.

The Beehive Star Cluster

A mysterious fuzzy patch of light has been noticed in the zodiac constellation Cancer since earliest human times. In ancient history, it was known as the Little Mist or Little Cloud, and also as Praesepe. This latter name means "manger," because two nearby stars were imagined to be asses coming to feed at it. Not until the invention of the telescope was the true nature of Praesepe determined: it is actually a marvelous cluster of stars faint enough and close together enough to blend into a patch of light to most human eyes under typical sky conditions. The cluster was the forty-fourth object on Charles Messier's famous list of deep-sky objects and thus designated M44. But in modern times the view of it through binoculars and small wide-field telescopes has earned it a popular nickname: the Beehive.

Many of M44's stars are of similar brightness and seem bunched or paired—truly like bees around their hive. But you must use a low-enough magnification to take in the whole cluster and keep its appearance concentrated.

M44 is big. It is about 1½° across and very close to the ecliptic, the midline of the zodiac, so it offers a large and in-the-way target for the Moon and planets to pass through. Bright planets overwhelm the naked-eye sight of M44 when they go through it, but a little optical aid beautifully shows the stars twinkling all around their brilliant visitor. Slower planets may linger in or near the cluster for a number of nights. The most highly reflective of large asteroids—44 Nysa—occasionally passes through M44. And in the past few decades, as luck would have it, two fairly bright naked-eye comets have passed quite near the cluster.

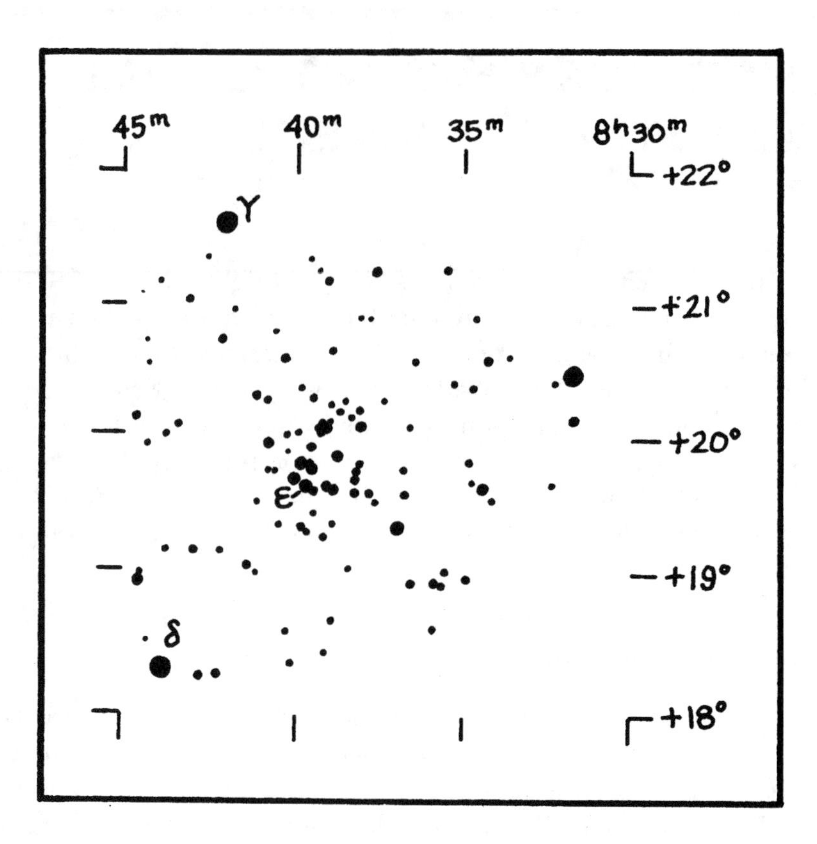

M44, the Beehive star cluster (center).

Ancient astronomical writers such as Aratos and Pliny noted a belief that when the stars were visible and Praesepe was not, bad weather was coming. This actually makes sense for naked-eye observers. For even though the magnitude 3.1 cluster has a greater total brightness than any star in dim Cancer the Crab, the brightness is spread out over a much larger area. The fact that M44 is an "extended" object means that it is more susceptible than the points of light that are stars to being washed out by light pollution—or by even a slight amount of cirrus clouds, which often precede rainy weather.

On the opposite end of the spectrum from being unable to see M44 with the naked eye is being able to glimpse a few individual stars in it with the naked eye. There are certainly a few sharp-eyed individuals who can do this under very excellent sky conditions. Most people are probably like me on the clearest and darkest nights: capable of seeing a beautiful unevenness of radiance across M44, and tantalized with a hint of glint from within it every now and then. How bright are its brightest stars? Three are slightly brighter than the traditional naked-eye magnitude limit of 6.5, eleven are brighter than 6.9, and fifteen are brighter than 7.5. But the proximity of them all to one another makes glimpsing individual members with the naked eye a challenge.

Other Clusters

Two other clusters worth note here are larger than M44: the Coma Star Cluster and the Alpha Persei Cluster (or, more properly, the Alpha Persei Association). The former is the third closest star cluster (after the Ursa Major cluster and the Hyades), situated just 290 light-years from Earth. The latter is located about 600 light-years from Earth, perhaps slightly farther from us than the Beehive.

The Coma Star Cluster is a charming scattering of 5th- and 6th-magnitude stars about midway between Beta Leonis and Alpha Canum Venticorum. It stretches across about 5° of sky, so a dark sky and either the naked-eye or fairly wide-field binoculars are needed to take it all in and make it look like any kind of concentration at all.

Yet ancient observers were fond of this cluster. In ancient Greece, they saw it first as the tuft of hair at the end of Leo the Lion's tail and then as the cut-off tresses of Queen Berenice. Berenice was a real-life queen of the Ptolemaic dynasty in Egypt, and the story goes that she promised to clip off her amber locks and place them in the temple if the gods would bring her husband home from war safely. He did return safely and she kept her word—but almost immediately the locks of hair disappeared from the temple. What had happened to them? The court astronomer said that the gods had transported them to the heavens, for look, there they were!—a beautifully disheveled smattering of stars high in the spring sky. This legend led to the surrounding constellation becoming known (today, officially) as Coma Berenices—Berenice's Hair.

The Alpha Persei group of stars is considered not really a cluster but a more loosely bound *association* of young stars. It is dominated by the 2nd-magnitude Alpha Persei, also known as Mirfak. Other members include moderately bright naked-eye stars up to 5° away in Perseus, but what strikes the eye as a cluster is a richer congregation of fainter naked-eye stars much closer to Alpha Persei. To be seen as a naked-eye sight, this cluster calls for fairly dark skies. But in binoculars or a rich-field telescope, the dim stars huddled near Mirfak shine out brilliantly all across the field of view.

Are there other open clusters that are best seen with such extremely low magnification? None are as close to us in space and therefore as spread out in the sky as the Ursa Major Cluster, the Hyades, the Coma Cluster, the Pleiades, the Beehive, and the Alpha Persei Cluster (to name them in order of increasing distance from us). But, of course, we could continue the progression out with slightly more distant and slightly less spread-out clusters—for instance, the giant M7, which glows about 900 light-years away from us near the tail of Scorpius. But we have to draw the line somewhere. We will consider M7 and other, more distant clusters—ones that are really at their best with at least slightly higher magnification in a telescope—in Sight 47.

Sight 28 Orion's Belt and Sword

It's impossible to avoid seeing the Belt of Orion when you are out on a winter night. Even people who have never learned the Big Dipper or remembered any other pattern of stars they ever saw will tell you or ask you about this: a seemingly perfect row of three equally bright stars, which they think maybe they saw in winter—the Belt of Orion.

Not only is it impossible to avoid the Belt when you are outdoors but also it has been impossible to avoid mentioning the Belt already in this book. We visited it briefly but prominently in Sights 9 and 16. But now we will look at the Belt of Orion more closely. With it we will examine the dimmer but strange and wonder-bearing asterism that accompanies it: the Sword of Orion.

The Three-Star Sentinel

Of all compact star patterns, only the concentrated little flock or hoard of the Pleiades can compete with the Belt of Orion in distinctiveness. But the Belt is brighter and bolder (though the Pleiades are prettier and more mysterious). When you first see the Pleiades—or see it again with fresh wonder—part of you asks yourself what this strange appearance could be. But when you gaze upon the Belt of Orion, you know it is a line of stars. The bewildering, astonishing question is: how could such a formation occur naturally? I hope no one supposes that extraterrestrial beings formed the Belt of Orion to be some kind of signal! Some of us will remember that with thousands of naked-eye stars in the sky, the odds favor at least a few being distributed in elegant and nearly perfect geometric forms. But it's still hard to get over the shock of actually seeing such an arrangement in the sky.

Even more marvelous is where this arrangement happens to be in relation to other bright stars in the pattern of Orion. In Sight 16, I discussed what I called the several "centralities" of Orion. But in this centrally located constellation, the center of centers is the Belt. The Belt is not only the center of Orion's pattern but also is located right at the celestial equator. That means

Orion's Sword, featuring the Orion Nebula.

it passes overhead at Earth's equator and should be visible from anywhere on our planet (though for practical purposes probably too near the horizon to be seen from the South Pole and certainly the North Pole).

When I was young, my brother told me that an imaginative friend of his had a special name for the Belt of Orion: the "Three-Star Sentinel." The name has always made a lot of sense to me. I'm not the only person who has had the fancy of imagining Orion as a whole being a noble guardian of the world (try listening to Ian Anderson's song "Orion" on Jethro Tull's album *Stormwatch*). But we can also imagine the three stars of the Belt, standing as they do virtually right on the celestial equator, keeping a watchful, protective vigilance over the world.

Alnitak, Alnilam, and Mintaka

When Orion is high up for an observer in Earth's northern hemisphere, the three stars of the Belt are, from lower left to upper right: Zeta Orionis, Epsilon Orionis, and Delta Orionis. (An acronym for mnemonic purposes here would be ZED—the British name of the last letter in the alphabet, Z.) But these three stars are also known by the proper names Alnitak (Zeta Orionis), Alnilam (Epsilon Orionis), and Mintaka (Delta Orionis).

The magnitudes of the three stars are 1.7 (Alnitak), 1.7 (Alnilam), and 2.2 (Mintaka). If Mintaka is the third star the eye usually comes to (in what is, for most people in Western culture, the natural reading direction, left to right), the eye can also determine that it is very slightly out of line with the first two. These slight departures of brightness and geometry from perfection are seldom noticed, however. The eye-mind system basically registers a straight line of equally bright, equally spaced stars in a row.

Each of these stars has its own special attractions to look for in binoculars or a telescope.

Alnitak has exciting nebulosity near it, elusive but sometimes visible in telescopes. The dimly glowing Flame Nebula to its northeast can be seen well in 10-inch and larger telescopes, but medium-sized telescopes can provide a glimpse once you know exactly where and what to look for. Then there is the streamer of dim nebulosity that extends south from Alnitak and has in it a tiny dark indentation that is one of the most famous of all photographic wonders of the heavens: the Horsehead Nebula. Amateur astronomers with 10-inch and larger telescopes have in recent years started glimpsing this **dark nebula** with the help of hydrogen beta filters. Over the years, a few of us have succeeded without filters and with smaller telescopes. But there's no denying that the Horsehead Nebula is a challenging visual object even for skilled observers under dark skies. To see the Horsehead or even the Flame Nebula requires keeping the blazing Alnitak out of the field of view, of course.

When Alnitak *is* in your field of view, try to see if you can detect its bright (magnitude 4.0) but close (2.5") companion star with fairly high magnification in your telescope.

Mintaka is actually slightly variable (by less than 0.2 magnitude in a period of 5.7325 days). It is a wide double star (53", magnitudes 2.2 and 6.8). What special attraction is associated with the middle star, Alnilam? There is a large (about 1½°-long) oval loop of stars around it, beautifully seen in a wide-field telescope or strong binoculars. Less obvious and rougher patterns of this kind can be traced around and near the other Belt stars, especially Mintaka, and there is a reason for this: the three Belt stars are actually part of a seldom-mentioned cluster, Collinder 70. The cluster consists of about 100 stars (brighter than 10th magnitude) spread across about 3° of sky. The total magnitude of the cluster, including the three Belt stars, is 0.4.

Wonders of the Sword

A few degrees south of the Belt shines an asterism that is quite different, but even more thrilling—if you have optical aid. I am referring to the Sword of Orion.

The Sword is quite different from the Belt. It is vertical (runs north to south) when Orion is upright and the Belt is at a diagonal. Its stars are not of about the same brightness but instead of steadily increasing brightness as one's eye traces it south. It is, overall, far less bright than the Belt—really only stars of modest naked-eye brightness, in fact. And yet it holds along its glittering length a number of telescopic wonders—and one marvel that is generally regarded as the most spectacular of all objects beyond our solar system when viewed in an amateur telescope.

The peerless wonder in the Sword is M42, the Great Orion Nebula. M42 is so grand an object it gets its own chapter—Sight 46. Here, we will concentrate on the other splendid telescopic sights of the Sword—beautiful sights that even experienced observers often overlook because they are mesmerized by the Orion Nebula.

Actually, the Orion Nebula—as M42 is most often called—includes a detached section that Charles Messier gave as a separate entry on his list. We now call it M43. But that is best discussed in our chapter on M42. If we ask ourselves what is the most prominent nebula in Orion other than M42/M43, the answer is another nebula in the Sword only 0.5° north of M42: NGC 1977. It surrounds the magnitude 4.1 total-brightness pair of 4.6 and 5.2 magnitude 42 Orionis and 46 Orionis, which are 4' 12" apart. Just north of this pair and NGC 1977 is the northernmost part of the vertical Sword: the 25'-wide, magnitude 4.5 open cluster NGC 1981. This cluster really does look like a sword handle to me in a telescope!

Be sure to admire the overall view of the Sword in good binoculars or a telescope with a very wide field. You'll see the cluster NGC 1981 at top (north end) and 42/46 Orionis surrounded by NGC 1977 as the second object down. Third in line is M42/M43—surrounding a double star, Theta Orionis, whose western member (Theta-2 Orionis) turns out to be the sky's most amazing quadruple star when looked at with enough magnification in a fair-sized telescope. (We'll examine Theta-2 Orionis, which lies at the very heart of the Great Orion Nebula, in Sight 46.) But wait—there is one more glitter in the Sword, the glitter that marks its point. That glitter is mostly supplied by the brightest star of the Sword: magnitude 2.8 Iota Orionis.

But Iota Orionis is not a solo jewel. First, it lies 8' from a dimmer but delightful double, Struve 745. And more magnification reveals that Iota itself is a marvelous triple star. Iota consists of a magnitude 2.8 primary with a magnitude 6.9 companion 11.3" away. A third, 10th-magnitude star is 49" slightly south and due east from the primary. One observer described the hues of these three stars of Iota as white, pale blue, and grape red. Another called the same stars "whitish sapphire," pale blue or aquamarine, and "pale mango." Struve 747 is a 4.8 and 5.7 magnitude pair a comfortable 36" apart.

But all of this is only half the story about Iota Orionis and Struve 745. Iota is actually the outstandingly bright member of the cluster NGC 1980. And in a dark sky with a good fair-sized telescope, you can see nebulosity around both Iota and Struve 745. As a matter of fact, in very dark skies you can see that there is a farthest wondrous loop of the Great Orion Nebula that extends all the way to Iota.

Still More Wonders of the Belt and the Sword

If you are a beginner who doesn't have a telescope or who just bought one, then reading the previous paragraphs may have your head swimming. How will you ever be able to track down all those double and triple stars, clusters, and nebulae? You shouldn't worry. When you first turn a telescope on the Belt and the Sword, you should simply enjoy the staggering riches that your eye beholds. You don't have to learn what each one is right away (or even anytime soon). As a matter of fact, no one has given more than a catalog number to some of the lesser but still delightful sights here. Indeed, the region is so rich that it's not just multiple star systems and clusters that catch the eye but luxurious random arrangements of stars. And there are the larger visual complexes you can form for yourself by your own favorite groupings of multiple stars, clusters, and invented asterisms.

Having said all this about just running your gaze in wide-eyed wonder around the telescopic (and binocular) fields of the Belt and the Sword, I must nevertheless persist in naming a few more stunning double stars that are in the vicinity of these star patterns if not quite in the patterns themselves.

Sigma Orionis is located up near Alnitak (the left star of the Belt) and is even more amazing as a multiple star than Iota is. The components are magnitudes 4.0, 10.3, 7.5, and 6.5 with generous gaps of 11", 13", and 43" between them. And 210" west of Sigma is Struve 761, which is a long triangle formed by the magnitude 8.0 and 9.0 stars 8' apart and about 68" from a magnitude 8.5 star. Finally, forming a nearly right triangle with the Belt to the southwest of it is Eta Orionis. It is a tight double, requiring a good telescope and steady night. Eta is a magnitude 4.0 star (actually slightly variable) that is just 1.7" from a 4.9 magnitude companion.

Are you in awe of the number and richness of wonders in Orion's Belt and Sword? Just remember, I have purposely excluded from this chapter any detailed consideration of the subject of Sight 46—the supreme wonder of the Sword, the Great Orion Nebula. Most amateur astronomers will spend more time on that nebula and its internal beauties than on all the other wonders of the Belt and the Sword combined!

Algol, Mira, and Other Dramatic Variable Stars

Sight 29

What is more constant than the stars? Poets and lovers have pledged their constancy by comparing it to the eternally fixed, eternally unchanging stars in the sky.

They really weren't that far from wrong. Of course, modern science has learned that all stars do change. We know from identifying stars at different stages in life that every star is born, grows old, and eventually dies. But these changes generally require millions or billions of years—periods of time so vast that they utterly dwarf the duration of human lives or even human history. So the stars *are* amazingly constant by human standards—except for **variable stars**.

Many thousands of stars are known to vary in brightness, and perhaps all stars vary noticeably in brightness at some time or other in their lives. The changes can occur over the course of minutes (even fractions of a second for pulsars), hours, days, or years. They can be as regular as an accurate clock or irregular beyond any hope of prediction. They can fluctuate so slightly in light output that only photometric studies can reveal the changes, not the human eye. Or they can flare or fade to thousands of times brighter or dimmer.

Most of the drama of variable stars is slow and requires careful scrutiny. So it's not surprising that many amateur astronomers find "faint fuzzies"—nebulae and galaxies—far more interesting than variable stars. But it is notable that the two greatest deep-sky writers of the twentieth century—William Tyler Olcott and Walter Scott Houston—were both absolutely devoted variable star observers. In fact, Olcott—the author of *Field Book of the Skies,* the classic stargazer's guide of the first half of the twentieth century—helped found and was the leading light of the AAVSO (American Association of Variable Star Observers). In the almost 100 years since its inception, the AAVSO has gathered more useful data than any other organization of amateurs of any branch of science in the world. If you think variable stars may interest you at all, you should go to the AAVSO Web site (www.aavso.org) and check out its marvelous record and selection of charts. These charts not only help you identify a particular variable star but also give you the magnitudes of *comparison stars*—stars of steady brightness with which to compare the variable star at different points in its *light curve.* The light curve is the curve on a diagram displaying the star's changes in brightness as a function of time—often over the course of a fixed *period* (amount of time from one maximum in brightness to the next).

Information about variable stars in general is useful, but one of the best ways to convey the excitement of these changing suns is to describe the brightest objects of the several most important classes of variable stars.

Algol the Demon Star

One night you observe the bright constellation Perseus, enjoying its major stars and taking note of their relative prominence. The next night Perseus is again one of the objects of your attention, and you trace the now-familiar form knowledgeably. Then good weather holds one night longer, and you are out again observing. You turn to Perseus, expecting a comfortable meeting of your expectations, but something is wrong. What is it? Now you realize: the second brightest star in Perseus seems to have lost half its luster. It is now outshined by five other stars in the constellation. What's going on? Is this radical diminishment permanent? Your question is answered over the course of the next few hours. The star is steadily kindling, as if by magic. And by halfway through the night it is back to normal.

What you've just seen was the minimum light and then return from eclipse of Beta Persei—a star far better known as Algol.

Actually, the little scenario I've just described—of a person who knew nothing about Algol being shocked by one of its eclipses—is not that common. Although Algol dims from magnitude 2.1 to 3.4 and back over the course of a 10-hour period every 2 days, 20 hours, 48 minutes, and 56 seconds, the eclipse will often occur during daylight or when the constellation is below the horizon—or on the night when you have bad weather. Amazingly, the variability of Algol was not discovered until around 1667 by Geminiano Montanari (the regularity of the period was not established until 1792 by John Goodricke). And yet one must be suspicious that ancient writers did know about the variable star. The ancient Greeks chose to picture it as marking the cut-off head that Perseus is carrying: the head of the monstrous Gorgon sister, Medusa, which even after death could turn anyone who looked at it to stone. Is this merely a coincidence or did some of the Greeks know that this star behaved in a spooky fashion? The medieval Arabs named the star Algol, "the ghoul"—a terrifying eater of human corpses. But it is possible that they were merely adapting the Greek version with a terror from their own tradition.

Algol is also known as the Demon Star. Whatever legends may say, however, there is a scientific explanation for Algol making its deep 10-hour-long wink at us every third day. Algol is the brightest of the class of variable stars known as **eclipsing binaries**. These systems feature two stars too close together to be separated by any telescope on Earth. But spectroscopic study of the point of light we call Algol helped us figure out that one star must be blocking the other from our view at three-day intervals. In this case, it is only a partial eclipse, and the major drop of apparent brightness occurs when the dimmer sun is passing in front of the brighter. (From our vantage, the brighter sun also passes in front of the dimmer at one point in their orbits around each other, but the drop in total brightness then is too slight to notice visually.)

Mira the Wonderful

Whereas stars that are eclipsing binaries dim because one component sun hides the other, there are many variable stars that change brightness because of actual physical changes in a single star. The most common and easy to see of these *intrinsic variables* are ones whose light fluctuates—irregularly or semiregularly—over long periods of time: the *long-period variables.* And the most famous of these is unquestionably Omicron Ceti, better known as Mira ("the wonderful").

Mira is located in the neck of Cetus the Whale. Most of the time, however, you won't see a star in this position if you are looking with just your naked eye because Mira is usually too faint. For a few months each year, however, the star is bright enough to see with unaided vision. Most years, it brightens to a peak magnitude of 4.0 or perhaps 3.0—and some years Mira gets even brighter. Several times in my life, I've seen this star outshine magnitude 2.5 Alpha Ceti and even rival the magnitude 2.0 Beta Ceti. The great thrill is that you never know just how bright it will get. At least once, back in 1779, Mira swelled in brightness to almost match the 1st-magnitude Aldebaran!

At minimum light, Mira is a telescopic object, often not much brighter than magnitude 10. But then comes a little more than three to four months of brightening, which carries it to maximum brilliance. The period between one of its peaks in brightness and the next is about 332 days, though the duration varies a bit. Notice that this figure is about 11 months. This means that Mira is brightest about one month earlier each year. That's a problem in certain years because the maximum then coincides with when the Sun is near Cetus in the sky. In 2007, Mira's peak occurs roughly in late February to early March and is just visible low in the west-southwest as evening twilight fades. The years after that should improve until around 2012, when Mira reaches peak brightness in early autumn (in the northern hemisphere) and rises around sunset to be visible all night long.

Mira is a semiregular variable, but some other stars are irregular variables. Both kinds of long-period variables are going through transitional periods in their lives, and their changes in brightness are associated with pulsations in which the size of these stars also changes. They are typically huge, cool red giants—Mira is one of those so cool that ordinary steam exists in it.

Other Kinds of Variable Stars

Algol is the most outstanding eclipsing binary and Mira is the most outstanding long-period variable. But there are many other kinds of variable stars.

Cepheids are pulsating stars that undergo their brightness variations with remarkable regularity. They exhibit a *period-luminosity relation*: the longer the period, the more luminous the star. By knowing this relation, astronomers have been able to calculate how distant these stars are. Observing Cepheids in nearby galaxies was a key to estimating the distance of the galaxies and understanding what galaxies are and the scale of the universe. The star that the class is named after, Delta Cephei, goes from one maximum to the next in a period of precisely 5.366341 days. During this period, its rise to maximum requires about 1.5 days, and the fall to minimum about 4 days. Its range is from magnitude 3.4 to 4.2. It is part of a little triangle at the southeastern corner of the house-shaped pattern of Cepheus. When brightest, Delta is similar in brightness to one other member of the triangle, Zeta Cephei; when faintest, Delta is similar in brightness to the final member of the triangle, Epsilon Cephei.

A variable star whose physical nature remains puzzling is a special kind of eclipsing binary: Beta Lyrae. *Lyrid type* or *Beta Lyrae stars* are believed to be pairs of stars so close together that their mutual gravity pulls them into elliptical forms that almost touch. There is little doubt that in the case of Beta Lyrae a giant streamer of gas flows from the larger to the smaller sun. The usual brightness of Beta Lyrae is 3.4, just a tiny bit dimmer than Gamma Lyrae. But the two unequally bright component stars of Beta each eclipse each other during a thirteen-day period. The first eclipse, about halfway through the period, is the lesser, dimming the overall magnitude only to 3.8. The brightness soon resumes to 3.4. But at the end of the thirteen-day period the dimmer star moves in front of the brighter and the brightness dips more deeply—to 4.1 (much dimmer than the comparison star Gamma Lyrae). Each of these eclipses elapse mostly within a two-day period, so the fall and rise of brightness is fairly rapid.

Some of the most exciting variable stars are those whose performances are rare and difficult to predict.

R Coronae Borealis type stars are sometimes also called *reverse novae.* R Coronae Borealis stays at about 6th magnitude for years—normally any pair of binoculars can show it quite prominently inside the cup shape of Corona Borealis the Northern Crown. But when R Coronae Borealis starts to dim, it drops to somewhere between 7th and 15th magnitude (average minimum around 12.5)—either still visible in binoculars or requiring something like a 10-inch telescope (and a very detailed finder chart) in dark skies.

Just outside the cup of the Coronae Borealis Northern Crown is a star that is the brightest of the class known as *recurrent novae.* This is T Coronae Borealis, but it is also known as "the Blaze Star." T's normal brightness is about 10th magnitude. In 1866, however, T shocked astronomers everywhere by flaring up to magnitude 2.0—brighter than Alpha Coronae Borealis (alternately known as Gemma or Alphecca). Unlike a true nova, however, T faded very rapidly—it was dimmer than the naked-eye limit in eight days and in less than a month back to its usual 10th-magnitude brightness. That seemed to be the end to a highly interesting story—until the night of February 9, 1946. Observers caught the Blaze Star at magnitude 3.2 but already fading, so that it's quite possible the star had gotten as bright as in 1866. It dimmed quickly, back down to 10th magnitude, just as before. Today, we can all hope to someday get an AstroAlert (see the sources section in the back of the book) to warn us before another outburst of T fades. But when Coronae Borealis is in your sky, you should always check the familiar semicircle to see for yourself if a "new" bright gem is shining not far from one end of it.

Even the Blaze Star gets "only" 8 magnitudes brighter over the course of its swift climb to peak light. True novae can increase their brightness by 10 magnitudes or considerably more. Supernovae can brighten by 20 magnitudes or more. These two fundamentally different kinds of exploding stars are the topic of our next chapter.

Sight 30 Novae, Supernovae, and Supernova Remnants

As we saw in the previous chapter, many stars undergo considerable changes in brightness, and these changes may occur regularly, semiregularly, or irregularly over a period of hours, days, or a few years. But novae and supernovae are far more brilliant, drastic, and unpredictable than any of these more common kinds of variable stars.

The mere idea that a particular supernova—one star—could for a few weeks outshine the combined light of a billion other stars in its galaxy is truly awesome. Imagine then seeing such a sight!

The only problem is that novae close enough to Earth to reach naked-eye brightness are infrequent, and the last properly observable supernova in our galaxy happened in 1604—more than 400 years ago! In a typical year, a dedicated amateur astronomer can learn about a few distant novae that are at least visible in medium or large backyard telescopes. And in many years, there will be a supernova in a galaxy tens of millions of light-years away that is nevertheless bright enough to detect in fairly large amateur telescopes. Part of the fun of observing remote novae is dreaming about the few closer, brighter novae that will occur in your lifetime and will temporarily change the naked-eye pattern of a constellation. Part of the fun of observing supernovae in other galaxies (which are extremely distant) is dreaming that just maybe yours will be the lifetime in which the long dry spell will end and another stupendous supernova in our own galaxy will brighten to rival or surpass the brightest planets in our sky.

Novae

Before the invention of the telescope, skywatchers occasionally noticed a star appear for a while where none had been visible before. Cultures had different names for them (the Chinese spoke of "guest stars"), but in Latin the term was *nova stella*, "new star," and then just *nova* for short.

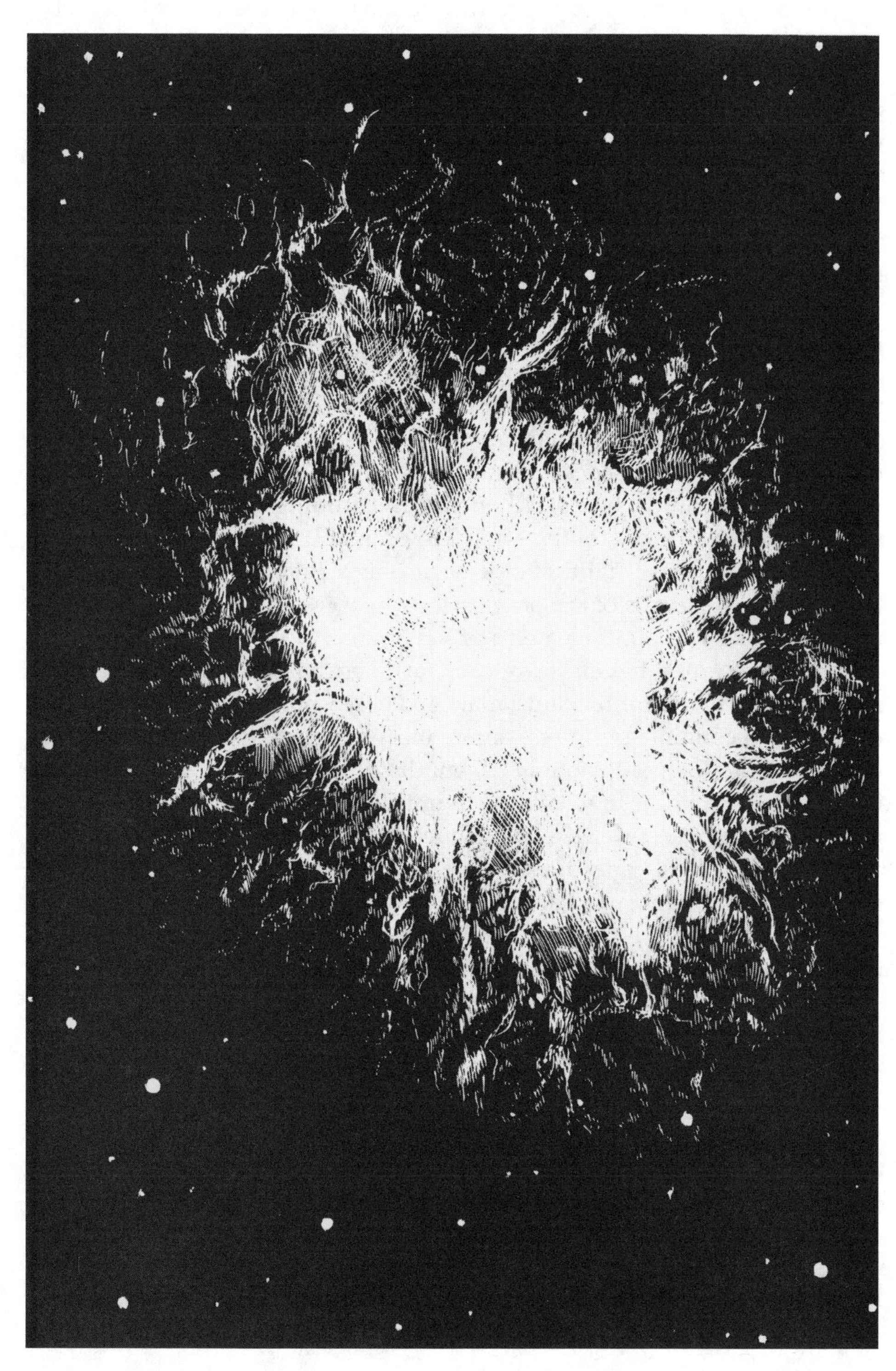

The Crab Nebula.

We now believe novae are usually produced when a main sequence star (similar in mass to the Sun) forms such a close pair with a white dwarf that material from the former starts pouring over the surface of the latter. An explosion results from runaway conversion of hydrogen to helium and the white dwarf may hurl off a few percent (maybe even 10 percent) of its mass and for a few weeks make the star system 10,000 or more times brighter.

Novae are most often found near the Milky Way band in the sky, for that is the more densely populated equatorial plane of our galaxy. The brightest nova in modern times reached magnitude –1.1 in Aquila the Eagle in 1918. The brightest in the last fifty years reached magnitude 2.2 in Cygnus the Swan in 1976. The very night before Nova Cygni 1976 got bright, I was walking with a friend at summer's end on a very clear evening. We walked about a mile out and a mile back under the stars with a stop at a dark field, where we stood admiring Cygnus in particular. My friend was not an astronomer and I remember him asking specifically about Cygnus. The next day I was off to college and, between bad weather and not finding out about the nova quickly, I missed the best of the event. How tantalizing in retrospect was that splendid pre-outburst night! Had it been one night later I could not have failed to have been one of the people who discovered on my own a new 2nd-magnitude star altering the pattern of the Swan.

Fortunately, over fifteen years later, in February 1992, communication and luck were better for me and I managed to see another nova in Cygnus at close to its maximum brightness—about magnitude 4.3. It was much dimmer than the earlier object I had missed, but the sight was still quite gratifying. I had seen several novae in telescopes in my life, but this was the only one easily visible to my naked eye. The experience of just standing with unaided eyes and surveying a landscape and skyscape with an exploding star in it was truly thrilling. And in binoculars and telescope Nova Cygni 1992 glowed a vivid orange. That's a color that one might think unexpected for an exploding ultrahot white dwarf but that certainly was exciting.

Since 1992, we have all entered into the age of the Internet. So now we have a good chance of learning about a nova as soon as it is discovered, perhaps even before it reaches maximum light. The best way to quickly learn about a nova is probably by receiving e-mail AstroAlerts (see the sources section in the back of the book). This is also the best way to learn as soon as possible about a new supernova.

Supernovae

It wasn't until the twentieth century that astronomers realized that what had always before been called novae really consisted of two fundamentally different kinds of events. One was the nova that we just described: an enormous but not necessarily fatal upheaval of a star located in our own galaxy and therefore close enough to reach moderate naked-eye brightness in some cases. The revelation came when astronomers ascertained that M31, the Andromeda Galaxy, was not mere thousands of light-years away. A "new star" named S Andromedae had appeared in M31 in August 1886 and brightened to reach magnitude 5.5. When it became clear that M31 was so distant—was in fact an entire galaxy far outside our own—astronomers realized that S Andromedae was no ordinary nova. It was an immensely more powerful event that became known as a "supernova."

Supernovae are typically about a million times brighter than novae. S Andromedae was so bright that it was visible to the naked eye from a distance we know now to be more than 2 million light-years. A supernova causes a star to grow millions of times brighter. There are several varieties, but the basic cause for the catastrophic event is the same: a star reaches a stage at which its mass and the resultant gravitation pulling it inward becomes too great to offset it by the outward radiation pressure from its nuclear burning. The star implodes, its core collapsing into a white dwarf, a neutron star or a black hole while hurling the rest of it—up to 90 percent of its previous mass—out into space. Unlike novae, which may happen again after 10,000 or 100,000 years, a star can go supernova only once; there is no return from the irrevocable catastrophe. But these explosions are credited with creating the heavier chemical elements and seeding the universe for the eventual production of planets with compositions like our own.

Amateur astronomers discover not only novae in our galaxy but also supernovae in other galaxies. All of us with even just medium-sized telescopes can hope to see a few supernovae in relatively close galaxies over the years. The closer the galaxy, the greater the apparent brightness of the object. You usually have to find out exactly where the supernova is located with reference to the center of the galaxy to make sure whether the particular 14th-magnitude, 12th-magnitude, or, rarely, even brighter point of light you're seeing is merely a foreground star in our own galaxy or is the real deal—an individual exploding star many millions of light-years from Earth. If the galaxy is one that is familiar to you, however, the identity of the supernova is quickly recognized. In any case, when you've ascertained which object is the supernova, an awesome realization comes over you: you are beholding a mighty event that actually took place long before any creature we could call human walked the Earth. It has taken millions of years for the light from that explosion to reach you.

The best supernova that most living astronomers—including myself—have seen was one in the great bright spiral galaxy M51. But people who lived far enough south or were able to journey there saw something much more spectacular in 1987: a supernova in one of the Milky Way's satellite galaxies. Supernova 1987A was easily visible to the naked eye. Those of us at northerly latitudes whose view of its light was blocked by the bulging solid body of the Earth were that one day showered by it with the subatomic particles called neutrinos—which pass through entire planets as if the worlds weren't even there.

But what every serious amateur astronomer wants to see is a supernova in our own galaxy. If such an event happened on the other side of the Milky Way and was dimmed tremendously by interstellar dust between it and us, it might appear no brighter than a bright nova to us. But there is a good chance it could look like the supernovae of 1604, 1572, 1054, or 1006. These objects shined, respectively, at magnitudes of about –3, –4, –6, and –8. In other words, they were about as bright as Jupiter, Venus, a thin crescent Moon, and a Moon approaching First Quarter.

That sounds impressive, especially when you consider that such objects stayed near peak for weeks, very bright for months, and visible to the naked eye for up to about two years. But most of the observational ramifications of having stars that bright are ones that wouldn't immediately occur to you and yet range from spectacular (at magnitude –3) to absolutely staggering (at magnitude –8). Planets such as Jupiter and Venus don't twinkle. Take a look at Sirius with a medium-sized telescope on a night when it is pulsing particularly vigorously. Imagine an object that bright to the naked eye hanging above a landscape, noticeably dimming other stars across much or all of the sky, shooting out dazzling bursts and darts of all colors, casting shadows, and even sending "shadow bands" (like those at a total solar eclipse—see Sight 2—but presumably much more prominent). Imagine as clouds passed in front of a Moon-bright supernova seeing the lovely cloud-corona disks of color we do with the Moon—only much more sharply defined and intense due to its being caused by a point of light. Whereas the large Moon struggles to be seen through thick horizon haze and air when it first peeks above the horizon, a supernova such as the one in 1006 would on a clear night truly burst across the landscape the moment it came above the rim of the horizon.

If you want to read more about supernovae in our galaxy and the stunning observational effects they create, you can check out the chapter titled "The Next Supernova" in my book *The Starry Room* (see sources section). How ironic it is that the last two really visible supernovae in our galaxy occurred just 32 years apart—and now we have waited just over 400 years without getting another. We don't have enough statistics about the frequency of these rare events in our galaxy to say that we are truly overdue. But the odds are improving. Let's all hang on as long as we can. There is some hope that many people reading this book will live to see the next supernova in our galaxy.

The Crab, the Veil, and Other Supernova Remnants

While we are waiting the long wait for the appearance of the next supernova in our galaxy, one of the things we can do is study the glowing remains of past grand explosions. There are only a few **supernova remnants (SNRs)** that are bright enough to enjoy in small telescopes. But these have an appearance unlike that of any ordinary diffuse nebula or planetary nebula, and there are few sights in astronomy as eerie and thought provoking.

Think, for instance, of the Crab Nebula in Taurus. This magnitude 8.4 patch of glow is conveniently located only about 1° northwest of Zeta Tauri, the star that marks the tip of the Bull's southern horn. The Crab is also known as M1—the very first object of Charles Messier's famous list. The nickname refers to the clawlike projections of this radiant patch's scalloped edges. You'll need a good 8-inch telescope to begin seeing that kind of detail in M1. But even if you're watching in a very small telescope, the reality of seeing this object will send a tremor through you. Why? The Crab Nebula, located about 6,500 light-years from Earth, is the remnant of a supernova that flamed up brighter than Venus in our skies a little less than a thousand years ago. It's possible that American Indian pictographs of a crescent and star in two different caves in northern Arizona are depictions of a conjunction of the Moon and supernova that was visible before dawn on July 5, 1054. Today, as this object shines eerily in your eyepiece, you can know that the star that survived the explosion is a *neutron star*—a star collapsed to the size of a city and now, in this case, spinning round at a rate of thirty times per second! Because of its angle of presentation to Earth, this object is not just a neutron star but a **pulsar**, emitting to us pulses of radio and other wavelengths of electromagnetic radiation—including very, very dim visual ones.

Different supernovae leave very different remains. Another easy SNR to observe is one from a star blast that must have occurred about 15,000 years ago. It may have been only 2,500 light-years away and thus lit up our sky like a thick crescent Moon. The remnant is now glowing strands here and there that define the outlines of a vast—3.5° by 2.7°—bubble. This is the Cygnus Loop, but the best known and easiest part of it is the strand of luminous filaments known as the Veil Nebula (NGC 6960). The Veil extends roughly north to south through 52 Cygni, a magnitude 4.2 star, that is about 3° south of magnitude 2.5 Epsilon Cygni, the easternmost star of the Northern Cross pattern of Cygnus. The Veil calls for low magnification—in dark skies it can be glimpsed even in 7×50 binoculars or a finderscope. But it needs a dark sky, and the sight of it in the narrow field of a typical medium-sized telescope can be elusive and ghostly. With larger amateur telescopes (and filters), the exquisite filamentary structure becomes prominent in this part of the

Cygnus Loop and a few others. NGC 6992 and 6995, the eastern sections of the Cygnus Loop, are 2.7° northeast of NGC 6960 and are actually brighter—they are just not blessed with an easy guide star for locating them. If you get a chance to look at any of these sections of the Cygnus Loop in a 16-inch or 20-inch telescope, grab it. With instruments of these sizes in a really dark sky, fantastical detail begins to show—in one place, the filaments give me the impression of being a luminous tornado.

Sight 31 STARRIEST FIELDS

If you're turning to this chapter right after reading the previous one, you may still be dreaming of what it would be like to see the brightest star that ever could light the nighttime sky—a relatively close supernova in our own galaxy. You may not live to see the day when so rare of an event occurs. While we are waiting, however, there is a heavens full of other wonderful sights that are available tonight. For instance, if you can't see the brightest possible star tonight, how about seeing the *most* possible stars. I mean seeing the greatest number of individual stars you possibly can in one naked-eye, binoculars, or especially telescopic field of view.

Starriest Naked-Eye Views

I don't think I've ever read anyone other than myself make a special point of discussing this question of where to find the starriest fields. So I must rely on my own experience, which like any individual person's, must be somewhat limited. But mine is especially limited in this inquiry by one important fact: I've never had the pleasure of being far enough south on this planet of ours to enjoy the full treasures of the southern circumpolar heavens. I suspect that even starrier views than I am about to recommend may exist in the far southern realms, especially in the constellations of Crux the Southern Cross and Centaurus the Centaur.

Here is what a Northerner thinks. The starriest area for really bright stars is surely Orion the Hunter (see Sight 16). But Scorpius has more stars brighter than 3.0 than any other constellation. And a look on a star map

A starry region of the Milky Way.

shows amazing crowds of moderately bright to dim naked-eye stars in the Milky Way regions of Puppis and Vela—though they are rather far south, and therefore somewhat low in the sky, for many Northerners to see well. The opposite season and height in the sky present a region that despite the distraction—wonderful distraction!—of its giant hazy bright star cloud, may offer Northerners the greatest number of stars visible to the naked eye. The region I mean is that of Cygnus the Swan.

Starriest Binoculars and Telescopic Views

I think that the competition for the starriest field in binoculars is far less complex and controversial. The naked-eye view depends so much on the size of the field we are allowing (a vast area to scan, a large one to try taking in with wide but fixed eyes, or a smaller area to focus on carefully) and on darkness and clarity of the sky. In contrast, the field allowed by binoculars ranges from about 10° (pairs with a little wider view than this are commercially available) down to about 4° (at which point rich-field telescopes with a larger aperture than all but extraordinary giant binoculars begin to take over). In this range of field widths, I believe the hands-down winner for the starriest fields is, again, parts of the constellation Cygnus. When your sky is so clear and dark that the naked-eye pattern of the Swan becomes lost and even the naked-eye view of Cygnus starts to look almost curdled with stars, the views in binoculars are among the most breathtaking sights in all of astronomy. Even under less than optimum sky conditions the visions you'll see here through binoculars are majestic.

What about the starriest view through a telescope? There are so many kinds of telescopes—so many apertures and different eyepieces and different field diameters. We can raise the question of whether we should count in this competition the richest open star clusters (see Sight 47) and the most resolvable globular star clusters (see Sight 48). But if we wish to talk about wide telescopic fields that are richly carpeted from wall to wall with individual stars, I would recommend looking at M24—the Small Sagittarius Star Cloud. A medium-sized telescope with the widest-field eyepiece can fit this entire 2°-by-1° brightest knot in the naked-eye Milky Way into a single field of view and show its individual stars down to 12th or 13th magnitude (or dimmer?). Not counting the ultraconcentrations of rich star clusters, is it ever possible to see more than a few hundred stars at a time in any single very wide naked-eye, 10°-wide binoculars, or 2°-wide telescopic field of view? Maybe not, but when you stare through the porthole of the spaceship that is your telescope at M24, you will gasp and feel your spirits soar as you yourself fall in, drown, come back up, and then proceed to revel and luxuriate in an ocean of stars.

THE SAGITTARIUS MILKY WAY REGION

Sight 32

This is the first of two chapters that are "tours" in this book. Is a tour really a sight? No, it is not. When faced with the richness of the heavens, I find that I must, after all, bend our rules a little. To get more of the great sights before us, I must remind us that our scheme of profiling the 50 best sights is a useful one, but that we shouldn't let it constrain us. Here's how I'll justify taking us on this telescopic tour of the Sagittarius Milky Way region: after you have taken the telescopic tour, go back and look over this entire region with the naked eye or the best parts of it with binoculars. Some of the objects we will visit on this tour are visible—even if as just fuzzy dots or patches—to the naked eye, and all are within the range of binoculars to see (though not in great detail). As you take in the entire region with your eye or binoculars, remember the close-up detailed visions of the individual clusters and nebulae you got through your telescope.

Getting Your Bearings

Indeed, the best way to get your bearing before starting the tour is to use your naked eye and binoculars to find first the guide stars and star clouds, then the brightest (naked-eye) nebulae and clusters before moving in on each one with the telescope.

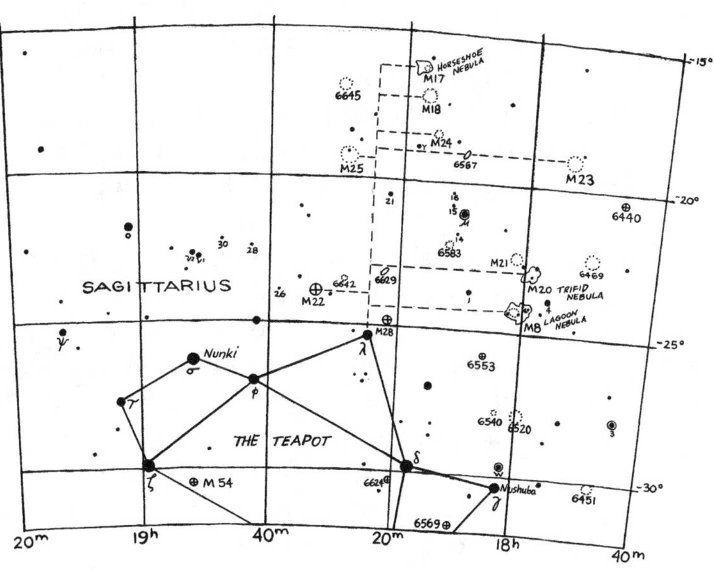

Locations of deep-sky objects in the Sagittarius region.

If you have a problem with haze or light pollution when you try this tour, you should start by locating the important stars of the constellation Sagittarius itself. You may recall that most of the brighter naked-eye stars of Sagittarius can be organized into the pattern known as the Teapot. It has two stars brighter than magnitude 2.5. The brightest is magnitude 1.8 Kaus Australis (Epsilon Sagittarii), which is situated in the lower right (southwestern) corner of the

Teapot (the bottom that is on the spout side). The second brightest is magnitude 2.0 Nunki (Sigma Sagittarii), which is situated in the upper part of the handle of the Teapot. The tip of the spout is marked by magnitude 3.0 Gamma Sagittarii. The top of the Teapot is magnitude 2.8 Kaus Borealis—much better known as Lambda Sagittarii. About half the width of your fist (or about one average binoculars field of view) to the northwest (upper right when these stars are highest) of Lambda Sagittarii is a dimmer star that is not part of the Teapot. This dimmer star, magnitude 3.8 Mu (μ on map) Sagittarii, is a very important one for helping to locate most of the best deep-sky objects of Sagittarius. Interestingly, while all the 2nd- and 3rd-magnitude stars that form the Teapot are found within a few hundred light-years of Earth, 4th-magnitude Mu Sagittarii is thought to be about 4,000 light-years away—it is a star that perhaps rivals Deneb and exceeds Rigel in true brightness.

There are just a few more stars we need to help us find our way to the many great deep-sky treasures of the Sagittarius Milky Way: the stars of the sting of Scorpius. It's easy to identify the sting itself because it is formed by that unmistakable pair of bright stars that has also been called the Cat's Eyes: magnitude 1.6 Shaula (Lambda Scorpii) and 2.7 Lesath (Upsilon Scorpii), just 35' apart.

Excursion to the Sting

We can begin our tour in earnest by scanning just a few degrees north and east from the two stars of the Scorpion's sting. There you cannot help but see, first in your finderscope (or binoculars), the two big open clusters M6 and M7. M7 is positively gigantic—a degree or more across—and glows at magnitude 3.3 or brighter. The famed observer and astronomy writer Steve O'Meara estimates M7 as magnitude 2.8. He calls it the brightest spot in the naked-eye Milky Way and says that to the unaided eye it looks like a great lost comet (without a tail). But remember that in a telescope this loose and sprawling cluster only looks superb at a very low power. Just 3.5° to the northwest of M7 is M6. M6 is dimmer and smaller—about magnitude 4.2 and 20' across—but still easily bright enough to see with the naked eye in fairly dark skies. Its greater concentration of stars makes it a better sight in most telescopes than M7. M6 is often called the "Butterfly Cluster" because its brightest stars form a stylistic outline of a butterfly. By the way, notice that M7 is right in the midst of a bright section of Milky Way, whereas M6—just those few degrees away—has a dark background (presumably because of dark Milky Way dust clouds behind it). M7 is only 780 light-years from Earth. Because M6 is about twice as distant, these two are actually rather similar in true size and brightness.

Perhaps our visit to M6 and M7 should be considered a separate brief excursion before we move over to Sagittarius for the main tour. I say this because all the remaining objects on our tour lie within a 10° circle of each other. It's that distant supergiant Mu Sagittarii that we referred to earlier that lies almost at the middle of that circle, with none of our wonders very close to it but all of them between about 4° to 6° from it. That's nice to know if you have binoculars, which most typically have a field about 4° to 6° wide (though I have two with fields *more* than 10° wide!). But perhaps the most efficient way to locate these deep-sky objects is with an equatorial telescope and a starting point at the star that marks the top of the Teapot: Lambda Sagittarii.

Clusters at the Offsets

It was a long time ago that I learned this trick of going a certain number of degrees due north from Lambda Sagittarii and then various distances due east or west to find each of the major deep-sky wonders in the Sagittarius Milky Way. I learned it as a kid in the late 1960s from several of the softbound books on making and using telescopes published by Edmund Scientific Company (now called Scientifics). Those books were both written and illustrated by Sam Brown. Some of those books are still available, and though much more has happened in the world of telescopes in recent years, they retain some of the best descriptions and illustrations of basic telescope information that you'll ever find.

In any case, if we begin at Lambda Sagittarii and go just 1.5° due north, then we only have to move the telescope along its polar axis 2° to bring us to the great globular cluster M22. Alternatively, it is easy for all but beginners to just scan 2.3° northeast from Lambda to get to M22. However you get there, the point is that you will want to stay. In Sight 48, I discuss bright globular clusters and hint that my favorite is the great M13 in Hercules. But I must confess that this feeling may be a result of affection for what is a more familiar sight from my early days: M13 passes high overhead for viewers at midnorthern latitudes and therefore is well placed for observation for much, much longer than M22. It is also far less dimmed by haze on most summer nights in the eastern United States. Nevertheless, when a summer night is crystal clear, I always turn to M22 with breathless anticipation. What I see then is probably the finest view of a globular cluster available from midnorthern latitudes. Rather than try to describe this magnitude 5.5 ball of radiance sparkling here, there, and everywhere with various numbers of its hundreds of thousands of stars, I'll do as Robert Burnham Jr. did in his famous handbook and refer the reader to J. R. R. Tolkien's description of a great gem called "The Arkenstone" in his novel *The Hobbit*: "It was as if a globe had been filled with moonlight and hung before them in a net woven of the glint of frosty stars."

Next on our northward journey on our equatorially mounted telescope's declination axis is a big, coarse, lovely, and underrated open cluster. This is M25, which you find by going just over 6° due north of Lambda Sagittarii and then offsetting just 1° to the east. M25 was somehow left out of the catalogs of William Herschel and John Hershel and therefore also got left out of the NGC—the New General Catalog. One of two bright stars near M25's center is a Cepheid variable star that fluctuates from magnitude 6.3 to 7.1 in a period of just over 6.7 days.

Nebulae (and More) at the Offsets

Now let's go even farther north—just over 9° north of Lambda Sagittarii. At an offset of just over 1° to the west shines M17, usually called the Omega or Horseshoe Nebula. This quite bright cloud of gas really does look like the Greek letter omega (Ω) or a horseshoe—or a few other things it has been nicknamed for (Checkmark Nebula and Swan Nebula). This fascinating object looks good even in rather small telescopes, though of course greater aperture brings out additional wonderful detail.

Let's leap north just a few degrees from M17, daring to leave our system of the line from Lambda and offsets to go just over the border from Sagittarius into Serpens. Here, we find M16—the Eagle Nebula and its associated star cluster. Actually, it is the star cluster that is more obvious in small telescopes, which will have trouble showing much of the nebula. But an 8-inch or 10-inch telescope and a nebula filter can start bringing out some wonderful detail. Robert Burnham Jr. looked at the spectacular professional photographs of this object and found in its form a new nickname for it: the Star-Queen Nebula. One of the first great and still most famous images from the Hubble Space Telescope shows a small part of the nebula looking like eerie human figures or fingers, etched with dark edges—columns sometimes called "pillars of creation," for new stars are forming in them.

If we go back to our north-south line from Lambda Sagittarii, we can take a trip to a star cloud and the star cluster in it. Going 7° due north from Lambda and then offsetting just over 2° west brings us to what I think may be the overall starriest telescopic field in the heavens: that of the 2°-by-1° Small Sagittarius Star Cloud, also known as M24. Actually, there used to be a lot of confusion about what the title M24 was being applied to: the star cloud or the little cluster within it. Within the star cloud is the open cluster NGC 6603, but the background is so rich that it takes a fairly sizable telescope to make out this slightly richer gathering.

Our final stops on the journey of offsets are those that take us to M8 and

M20. Actually, we should visit the open cluster M23, which is 6.5° north of Lambda and then over 7° west. It's a fine sight. But it must play second fiddle to the virtuoso objects M8 and M20: the Lagoon Nebula and the Trifid Nebula.

M8 and M20 are close enough together and conspicuously unique enough to require only one offset. Let it be, after a trip of just over 1° north from Lambda Sagittarii, an offset of 5.5° west—to M8, the Lagoon Nebula.

For observers at midnorthern latitudes, only winter's M42, the Great Orion Nebula (see Sight 46), is a mightier nebula than M8. Unlike M42, M8 is not overwhelmed by any bright star in it for naked-eye viewers. Indeed, M8, which shines at about magnitude 4.5 and is more than a degree wide in its longer dimension, is a sizable puff of light rather easily visible to unaided vision in a clear, fairly dark sky. When you turn a telescope on it, however, it becomes the amazing Lagoon Nebula. Even a fairly small telescope reveals the dark lagoon—really more like a channel—that separates the two bright halves of the cloud. It also shows the brightest visible star that is causing the nebula to shine on its own—the magnitude 6.0 star 9 Sagittarii—and the open cluster NGC 6530. A medium- to large-sized amateur telescope begins to show color and lots of detail—especially around the 30"-long brightest part of the nebula, near 9 Sagittarii. Because of its shape, this region is sometimes called "the Hourglass."

Centered just 1.5° north-northwest of M8 is another of the premier nebulae of the heavens: M20, the Trifid Nebula. M20 owes its nickname to the dark lanes that split its bright emission part into three sections. Nearby is the reflection nebula part of M20, which is shockingly blue on long-exposure photographs beside the intense red of the emission areas. These colors are at best only faintly visible in medium-sized telescopes. M20 contains a magnificent multiple star, and the open cluster M21 is prominent near the nebula.

Nebula Row

Maybe you won't use the offset method we've employed on this tour. Maybe you'll get lazy and just dial up all these objects on a computer-run go-to telescope. That makes sense if you are trying to save time, or sky conditions are less than optimum. But if you try looking for these wonderful objects for yourself, without computer assistance, you'll learn a lot and see many glorious sights along the way that are not in the telescope's memory.

Actually, if you have a Dobsonian or other kind of altazimuth mount, you won't be able to use the offset method described in this chapter (or at least not very easily). But you can achieve the same results, daringly and with a little more difficulty, by pure "starhopping." This involves first looking at a star map and seeing how many field widths of your telescope it is from a bright object to a dimmer one. Then you try to use slightly overlapping fields, one after the other (with guide sights from your map along the way), until you get to your target.

What is the best way to starhop to each of the wonders we've mentioned in this chapter? You can draw a nearly straight line only about 8.5° long from M17 (the Omega Nebula) through M24 (the Small Sagittarius Star Cloud) to M8 (the Lagoon Nebula)—and M16 (the Eagle Nebula or Star-Queen Nebula) and M20 (the Trifid Nebula) are easy to find near the ends of this line. Four of the five premier diffuse nebulae visible from midnorthern latitudes (winter's Orion Nebula is the fifth—and best) are visible—with nearby clusters and star cloud—along this one short, straight road. I call it Nebula Row.

Sight 33 The Great Andromeda Galaxy

Sometimes astronomy writers suggest that except for the Great Orion Nebula, the grandest of all deep-sky objects is M31, the Great Andromeda Galaxy. Yet at the same time, some beginning amateur astronomers express disappointment in their view of M31 through a telescope. The disappointment is no doubt partly a consequence of the novice's expectations being set so high—not only by words but also by spectacular photographs of M31. Still, I think a further reason for beginners' sometimes cool reactions to the Andromeda Galaxy is another and very interesting one. It is the fact that there is no other celestial sight whose visual grandeur is so much the *combination* of the very different appearances of it with different optical instruments. You really have to view M31 with the naked eye, binoculars, a low-power telescope, and a high-power telescope, and then put these visions all together in your mind to appreciate the full measure of visual glory that this majestic galaxy offers. Then, of course, you should learn or review what it is you are actually seeing. Learn the nature, orientation, and the immense scale—in both time and space—of the lovely sights you are seeing.

If you do all this, you really will find that M31 is almost as visually appealing as M42, the incredibly beautiful and complex Orion Nebula. And the Great Andromeda Galaxy is perhaps even more mentally appealing.

The Great Andromeda Galaxy.

M31 with the Naked Eye

M31 was noticed long before the invention of the telescope. It is recorded at least as far back as the tenth century, when it was called "the Little Cloud" by the Persian astronomy writer Abd al-Rahman al- Sufi.

The naked eye, looking straight overhead on the middle of a November evening from midnorthern latitudes, sees a strange elongated smear of soft radiance. Or at least it does if your sky is relatively dark. The elongated glow really draws the eye to itself in very dark skies. Otherwise, you can find it by going from the 2nd-magnitude Alpha Andromedae to the 2nd-magnitude Beta Andromedae and then making a right turn a few degrees to the magnitude 4.5 star Nu Andromedae—slightly beyond which is the magnitude 3.4 glow of M31.

You may need binoculars to assist you in viewing the elongated form of M31. But the maximum length might actually be visible, in extremely dark skies, not with binoculars but with the naked eye. That length is an utterly amazing 4° to 5°. What inevitably increases one's awe is the knowledge of how distant this object is and, even very roughly, what it is. M31 is located well over

2 million light-years from Earth. There is an expression, "On a clear day, you can see forever." Well, on a clear autumn night you can literally see more than 2 million light-years, which is the same as seeing more than 2 million years into the past. You're doing both those things when you are staring up at the Great Andromeda Galaxy. And what is this delicate yet ineradicable slip of seeming phosphorescence? It is our huge Milky Way Galaxy's much bigger sister, a congregation of hundreds of billions of stars more than 200,000 light-years across. The flame of several hundred billion stars reduced to a gentle flicker. And ineradicable? Well, *our* view of M31 is certainly capable of being eradicated—by an increase in light pollution even at your rural site. But M31 itself may endure longer than anything else your unaided eyes can see—in fact, maybe longer than our Milky Way if it rips our galaxy apart as some scientists think it will about 4 billion years from now. (More on that in a moment.)

The galaxy M33 might possibly be slightly farther than M31, and it certainly is within range of careful naked-eye observation under very dark sky conditions. But for most people in most places, M31 really is the farthest sight that can be seen with the unaided eye.

The Companion Galaxies, Dust Lanes, and Star Cloud

What more of M31 can one see with binoculars than with the naked eye? More of its structure and the beautiful sort of translucence of its bright central region. Even in the very small, inferior telescope of Galileo's observing rival Simon Marius, M31 was (in Marius's words, as translated by the pioneer and science writer Willy Ley) "like a candle flame seen through the horn window of a lanthern [lantern]."

M31 is so long and the light of its outer regions so spread out that one must use very low power to see them well. All too often a beginner with a small telescope uses too high a power and gets only a confusing sight of the middle part of the galaxy filling the view, bright but diffuse and hard for the uninformed (especially the uninformed with a poorly collimated telescope) to fathom.

Let's back up a bit. When we see one of the famous long-exposure photographs of the Andromeda Galaxy, what are we seeing? A spiral galaxy—kind of like a pinwheel—we are told. But this pinwheel is tilted just 13° from edge on. So the spiral arms are primarily indicated by the presence of several dark streaks that are a partial view of the areas of dark gas and dust between the bright, starry arms.

Now we are looking at M31 in a low-magnification view through a fairly large amateur telescope. We do see one or two of the dark dust lanes—the most prominent is curved to the northwest of M31's center. We may have to move our field of view around a bit, but much more easily than the dark lanes we can see two bright patches to either side of M31: its two most prominent satellite galaxies. These two objects are M32 and M110. M110 (until recent decades known only as NGC 205) is an 8th-magnitude elongated ellipse of glow about 35' northwest of the center of M31. M32 is similar in brightness to M110 but closer to the glow of M31—only 24' south of M31's center. It is also rounder and smaller than M110; in fact, at very low magnification M32 can be mistaken for a fuzzy, out-of-focus star (especially when atmospheric seeing is poor or your telescope is not in good collimation). These two elliptical companion galaxies of spiral M31 are delightful sights south (M32) and northwest (M110) of the great spiral. But near the southwestern edge of the bright extent of M31 that is visible at a medium telescopic magnification, there is a patch in the spiral arms that is brighter than any other: NGC 206. It is not another satellite galaxy but rather a vast star cloud in M31. By the way, the west side of M31 is closer to us, and this stirs a remarkable thought: M31 is so large that the light from the western part of it closer to us must reach us 100,000 or even 200,000 years sooner than light from the eastern part, which is farther from us.

The Nucleus and the Majesty

What about the central regions of M31? They are a blaze—but a gauze-softened blaze!—of glory on any clear night. But on a night of good atmospheric seeing, when images are steady and crisp, up the power on your 8-inch or larger telescope. What do you see? Within the layers of the golden hub of M31's central region, you may see a nearly starlike "nucleus" of the galaxy. It actually measures about 2.5" by 1.5". It is nothing more than a particularly concentrated region of stars blended together into one glow. But what an amazing structure. This nucleus should be only about 50 light-years wide and yet may contain over 10 million stars. *Ten million stars!* That means fifty or sixty stars in each cubic light-year, collisions inevitable, a planetary resident's sky so star encrusted that night would seem a starry day. And yet do we really understand what is going on in this region? There is now evidence of not just one but two massive black holes at the heart of M31.

By the way, with large amateur telescopes many of M31's several hundred globular star clusters can be identified, and with apertures of 16 inches or more, the 17th-magnitude brightest blue giants in the spiral arms can start becoming individually visible.

The final touches of majesty are supplied to what we see of M31 by adding information about the size and place of this mighty spiral galaxy in the universe. M31 may be at least 1½ times larger and twice as massive as the Milky Way Galaxy. These two each rule one of the two subgroups of what we call the "Local Group" of galaxies—several dozen galaxies, mostly relatively small. The third largest member of the Local Group is the nearly face-on spiral M33 (see Sight 50 for more on it), but it is part of M31's subgroup. M33 lies only about 700,000 light-years from M31, so the latter must appear more than three times bigger and more than nine times brighter from M33 than it does to us. There is a suspicion, but no proof, of a streamer of gas connecting these two galaxies. But what is truly fascinating and beautiful is the relationship of M31 and our Milky Way. M31 is only about ten to fifteen of its own diameters away from us—and, by the way, we happen to be on the side of the Milky Way closer to M31 in this era of Earth's history. Our Milky Way takes over 200 million years to rotate once out at our solar system's distance from its center, and M31 may have a similar speed of rotation (or somewhat swifter?). But, like many pairs of large spiral galaxies in the universe, the Milky Way and M31 have similar tilts toward each other and rotate in opposite—that is, complementary—directions.

Their orientation to each other suggests that the Milky Way and M31 were born together with the Local Group. The two big galaxies—the Milky Way and M31—seem to be revolving around a common center of gravity (with their satellite galaxies circling them in periods of about 50 million years). But what will happen to them in the future is what's most amazing.

Most galaxies in the universe have their light *redshifted*—their wavelengths shifted toward the red end of the spectrum—because they are moving away. The cause of them moving away from us is the expansion of space itself that continues 14 billion years or so after the Big Bang. But M31's light is *blueshifted*, indicating it is heading toward us. Actually, this is not surprising considering that we might expect the Local Group of galaxies to be close enough together for their gravity to keep them together and for some members to be *temporarily* drifting slowly closer to each other. But some astronomers have calculated that M31 is not just drifting a little closer to us for a while. No, they tell us, the Great Andromeda Galaxy is going to collide with the Milky Way about 4 billion years from now. By then, our Sun will have swelled into a red giant, perhaps having already engulfed or almost engulfed a cinderlike Earth, and probably forcing us (whoever or whatever the residents of Earth may be so far in the future) to move to the outer solar system. But imagine, if you can, M31 looming vast, filling our entire sky. Then what would happen? Galaxies tend to sift right through one another—but not, of course, without tremendous disruptions. We might see supernovae going off like firecrackers in both galaxies' spiral arms, stupendous outbreaks of starbirth of new, hot, blue blazing stars being induced. The Milky Way might be slowly pulled apart like a big gooey cinnamon bun and stars scattered wholesale left and right. Astronomer William K. Hartmann points out that our Sun and its solar system might well be ejected—without the slightest jostling of the planets—out alone into the depths of intergalactic space.

Are these calculations accurate? Is our galaxy really fated to collide with M31? It's hard to say, but it is something astounding to reflect on as you observe M31. One thing, at least is certain: as you watch the Great Andromeda Galaxy, every second brings us 50 miles closer to it.

The Realm of the Galaxies

Sight 34

In chapter 32, we took a telescopic tour of the spectacular region of clusters, nebulae, and star clouds that decorate the Sagittarius Milky Way. In that chapter, I tried to justify the idea of calling a tour a sight. Whether I succeeded or not in that rationalization isn't important. What is important is whether I succeeded in making a tremendous wealth of wonderful celestial sights more easily accessible. That is again my goal as we take our second tour, this of the sky's richest of all clusters of galaxies for medium-sized telescopes: the Virgo Galaxy Cluster, also known as "the Realm of the Galaxies."

Before we start our tour, I want to stress that even the brightest galaxies of the Virgo Cluster are a challenge for beginning amateur astronomers. You'll need a fairly dark sky and at least a medium-sized telescope to identify these. In truth, a novice is better off starting with not just the previous chapter's very bright galaxy (M31, with its easy-to-find companion galaxies) but also with the especially bright galaxies of Sight 50. The reason that this tour of the Virgo Cluster occurs at this stage in the book is that it is a sight that is laid out over a fairly large area. The sections of this book, you will recall, are in order of the width of field needed for their sights—from widest (sights spread over the entire sky) to narrowest (sights concentrated into a very tiny high-magnification telescopic field of view). For the most part, this order makes a lot of sense (as explained in the introduction), but here I need to stress that novices may do better trying for the galaxies of Sight 50 before looking for the dimmer, though marvelously numerous galaxies of the Virgo Cluster.

Where and What

Draw a line from Epsilon Virginis (Vindemiatrix) to Beta Leonis (Denebola). Along and just north of the Virgo portion of this line is the central concentration of the Virgo Galaxy Cluster. Some of this concentration of galaxies spills over the boundary from northwestern Virgo into southern Coma Berenices (the constellation of Berenice's Hair). The Virgo Galaxy Cluster is thought to consist of about 2,500 galaxies and may be the core of a larger supergalaxy of 12,000 galaxies. Our own small Local Group of galaxies (see the previous chapter) may be on the fringes of this larger structure and revolving (presumably over the course of *billions* of years?) around the center we see there at the Virgo-Coma border. The distance to the Virgo Galaxy Cluster seems to be about 50 to 80 million light-years but may include some galaxies much closer to us such as M104 (the Sombrero Galaxy—see Sight 50).

What is certain for observers is that there are dozens of galaxies in the Virgo Cluster bright enough to detect even with a 4-inch telescope. To see most of them as more than indistinct little blurs of light, however, a good 8-inch telescope and a dark sky are recommended. With such an instrument and sky conditions, more than a hundred of these galaxies are visible within an area measuring only 12° by 10°. There are, within this area, individual fields of view that are richest of all. Place the galaxies M84 and M86 at the center of your low-power eyepiece's field and you may be able to see a total of ten galaxies at once. With a slightly wider field and M84 and M86 on the southwestern edge, you may be able to trace about a dozen galaxies in a zigzagging line called Markarian's Chain.

You can dial up these galaxies with setting circles and a list of their positions or have your go-to telescope's computer take you to them. But you will learn more and see so much more along the way if you discover these galaxies by taking certain paths through the sky and looking for certain memorable arrangements of the galaxies.

Entering and Organizing the Virgo Cluster

Our journey into the Realm of the Galaxies begins with magnitude 2.8 Epsilon Virginis and magnitude 4.9 Rho Virginis. Epsilon is easy to find and from it you must travel about 5° west to arrive at Rho Virginis—the only star of 5th-magnitude brightness in that position.

We encounter our first major Virgo Cluster galaxies just 1.5° north of Rho: magnitude 9.6 elliptical galaxy M59 and—just half a degree east of M59—magnitude 8.8 elliptical galaxy M60. Roughly 1° west and slightly north of

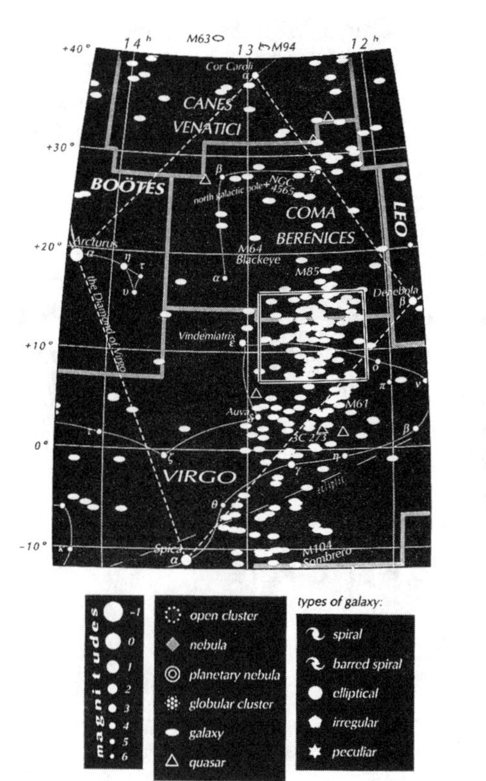

The Realm of the Galaxies.

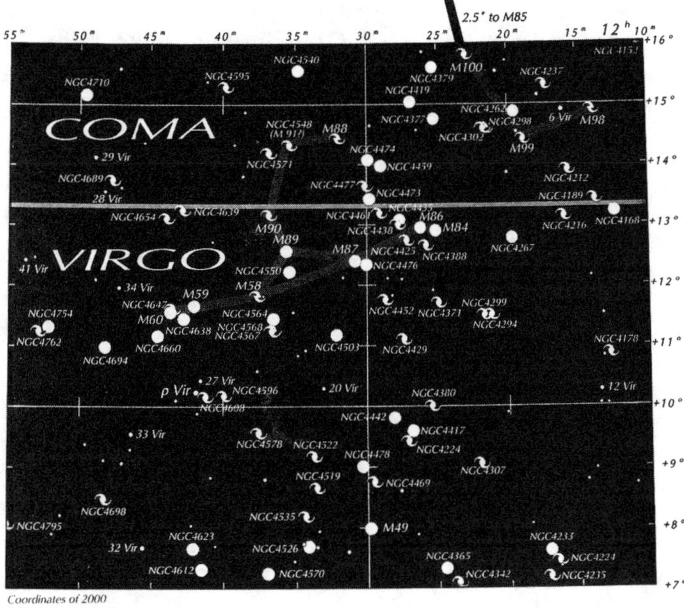

The core of the Virgo Galaxy Cluster.

M59 is yet another Messier galaxy: magnitude 9.6 spiral galaxy M58. M60 is one of the most massive galaxies known, perhaps six times as hefty as our big Milky Way, and it has a faint but observable companion galaxy (NGC 4647). But the fascinating trio of M60, M59, and M58 is just the start of a 5°-long straight line of six total galaxies that continues northwest with M87 and the pair M84 and M86. You can make the line a gentle arc and extend it another 4° northwest to M99 and M98 to increase the total to eight galaxies. But another plan can be tried. Forget M99 and M98 for the present. On the diagram above, notice that M58 and M87 form the base of a long, flat triangle with galaxy M89, which itself has galaxy M90 only 0.75° northeast of it. And then notice that you can complete a hook by curving to M88, just over the border in Coma Berenices. In doing so, you have, along with the M60 to M86/M84 line, constructed a "coat hanger" of galaxies. This coat hanger of galaxies is only about 5° long and 3° tall but includes ten Messier galaxies of the Virgo Cluster!

But the coat hanger of galaxies doesn't include M99 and M98. What should we do with them? They (both over the border in Coma) form a triangle with M100, and this triangle hangs on a stem from a yet more northerly galaxy, M85. The coat hanger and the "hanging triangle" together organize no less than fourteen galaxies for your ease of finding and delectation. By the way, the brightest of the fourteen in our sky and the most massive in outer space is the nearly round M87. It is one of the most massive of all galaxies known, containing about 800 billion times the Sun's mass—perhaps 3 billion Sun-masses just in a mighty central black hole alone!

The only certain Virgo Cluster galaxies that are Messier objects and not in our coat hanger and hanging triangle are M49 and M61. Magnitude 8.4 elliptical M49 is several degrees almost due south of M87. The smaller, dimmer M61 is a few degrees southwest of M49.

Still More Galaxies and the Galactic North Pole

Not every Virgo Cluster galaxy that is fairly bright was noticed by Charles Messier. There are nine galaxies in Virgo and Coma that are brighter than magnitude 10.0 but are not Messier objects, only NGC (New General Catalog) objects. Of course, not every one of these, just because they are in this section of the sky, is necessarily a true member of the Virgo Cluster. A good example is NGC 4565. It is located in Coma Berenices but is much closer to us than the Virgo Cluster galaxies—"only" about 20 million light-years away. NGC 4565 is possibly the most prominent of needle-thin edge-on galaxies in all the heavens. There is even one *Messier* galaxy in Coma Berenices that does not belong to the Virgo Cluster. It is M64, a magnitude 8.5 spiral galaxy with an obscuring dust cloud in it that earns it its nickname "the Black-Eye Galaxy."

If we go a little farther afield—into Ursa Major, for instance—we find even more galaxies—prominent galaxies—that are not members of the Virgo Cluster. Why are so many galaxies found in this general section of the heavens? Because this area is as far as possible from the obscuring dust and gas that lies in and near the equatorial plane of the Milky Way Galaxy. That *galactic equator* is marked by the softly glowing Milky Way band. The *north galactic pole* is the point in the heavens where we are looking straight "up" out of the equatorial plane of the Milky Way (that is, at a right angle to the galactic equator). Thus, it is as if we are gazing up out of a skylight and seeing in it unobstructed great numbers of galaxies—including the big Virgo Galaxy Cluster.

Field of View

1° to 0.1° or LESS

(MEDIUM TO NARROW TELESCOPIC FIELD)

Overall Telescopic Views of the Moon

Sight 35

There's no doubt about it: if you have an astronomical telescope and know some people who haven't really looked through one before, there is one easiest and just about foolproof way to get them to gasp. Show them the Moon.

To the naked eye, the Moon's face seems at a casual glance to show nothing more than a few curious gray markings on an otherwise uniformly bright surface. Actually, more careful examination with the naked eye can show considerable complexity in those markings and sometimes patches on the Moon that are brighter than average. The key to seeing the most detail with the unaided eye is to look around sunrise or sunset, when the sky is neither too bright nor too dark in comparison to the lunar surface. There is an additional kind of very meaningful detail that can be spotted with unaided vision: a jaggedness to the *terminator*—the line separating day and night on the Moon. Even as far back as ancient times, people shrewdly guessed that unevenness of the terminator must mean that there are appreciable variations in the height of topography on the Moon. High peaks catch sunrise long before the valleys near them and likewise remain illuminated after the Sun has set in the nearby valleys.

Still, even after we credit that the naked eye can detect more about the lunar surface than most people would ever guess, the fact remains: a view of the Moon through optical aid is a transformation so rich and complex and revelatory that it truly takes our breath away.

You will very much want to put in one of your higher-magnification eyepieces and look at the

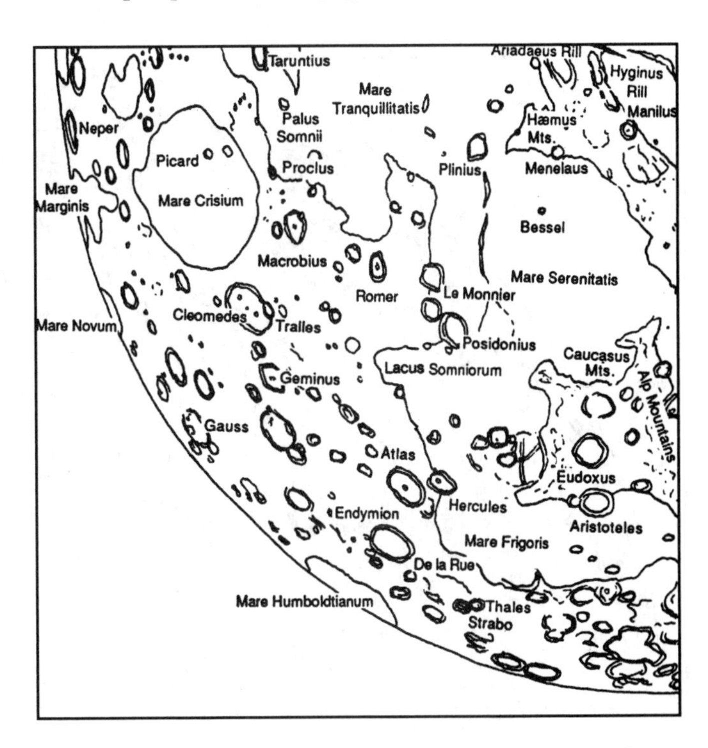

Moon Quadrant 1—the Northeast Quadrant. South is up and east is left, as in most telescopes.

exquisite fine detail of lunar features. But we will do that in the next chapter. For now, let's take a mostly low-magnification view. Among the many individual lunar sights, the view of the entire Moon (or at least a quadrant of it) filling your field of view has got to be the best of all. It is probably the most impressive of all astronomical sights that we can access night after night. It is also a living map that we should survey at considerable length before deciding the different places we want to focus more closely on with higher magnification in the next chapter.

Maria, Craters, Mountains, Rills, and More

Even steadily mounted binoculars can reveal several hundred features on the Moon. It has been estimated that a good 6-inch telescope in very steady "seeing" can show something like a million different lunar features. Where do we begin an exploration of this mind-boggling richness? First, with some general observations.

A small telescope shows that the gray smudges that the naked eye sees on the Moon are actually vast gray plains. People long ago wondered if these darker regions might actually be oceans. We still call these areas "seas" or, what is the same thing in Latin, *maria* (pronounced MAH-ree-ah; singular, mare [pronounced MAH-ray]). Scientists have figured out that the maria were formed early in the history of the solar system when asteroid-sized objects called *planetesimals* were common and bombarded the Moon. The immense basins that their impacts created quickly filled in with molten rock that then cooled, leaving the plains we now see.

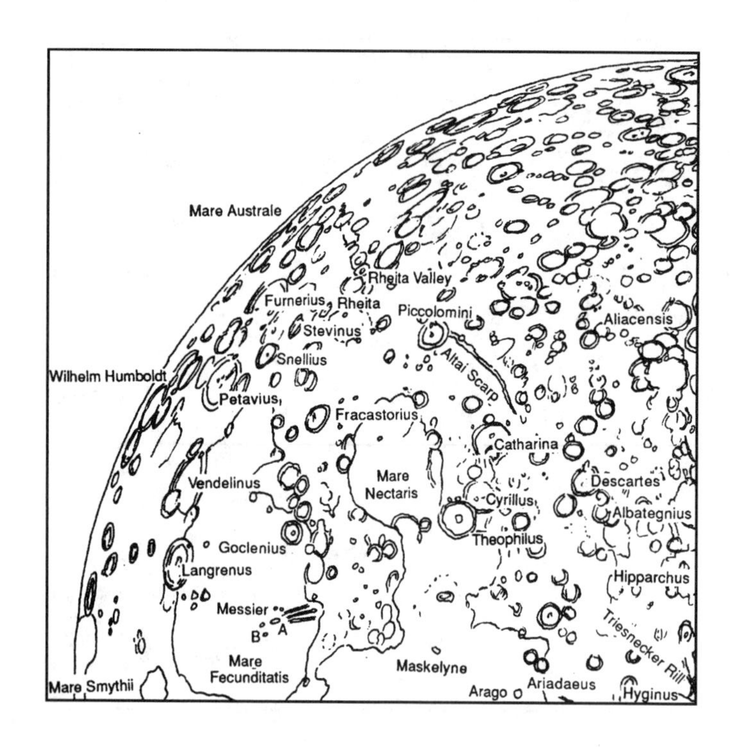

Moon Quadrant 2—the Southeast Quadrant. South is up and east is left, as in most telescopes.

As you scan the overall globe of the Moon more closely, even at a low magnification, you see many large craters and tremendous numbers of smaller ones at the edge of resolution. There is no mistaking a lunar *crater*—well, not usually, though there *are* some very old, worn, and partial craters that may be overlapped by fresher ones or half-buried by material from nearby maria. There are really several subcategories within the larger one "crater." But what about *mountains* and *mountain ranges*? The latter can often be detected forming a rugged border to part of a maria. Individual isolated lunar mountains are dramatic when the terminator is near them, and they cast long, sharp shadows. *Valleys* and *rills* are more or less narrow depressed areas, the latter being generally more like a narrower ravine—but actually these terms have been applied to a variety of strange lunar features, each of which needs to be explained and studied in its own right. *Faults* and *domes* generally require higher magnification, but in the former category the Straight Wall is a spectacular feature—under the proper lighting (see the next chapter).

Foreshortening, Lighting, and Rays

You may not automatically notice something about even the big maria if you are not alerted to it: lunar features are less rounded and more elongated the closer they are to the edge of the Moon. This is simply a matter of perspective. Scan near the middle of the Moon's disk and find a mare—or rather a crater, for craters are by nature round (maria tend to have some circular edges but are more irregular in shape). You'll see that the crater does appear nicely circular. But now scan closer to any edge of the Moon. The craters become more elongated ellipses, and finally odd dark lines at the lunar **limb** (edge), where we are looking at them at a very grazing angle around the side of the Moon.

Moon Quadrant 3—the Northwest Quadrant. South is up and east is left, as in most telescopes.

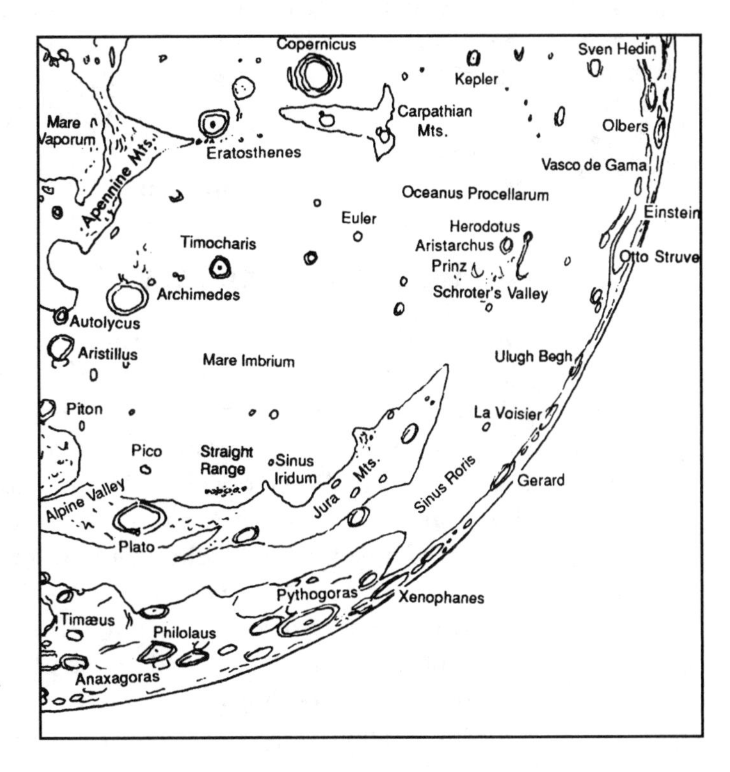

Now use your telescope at any phase other than a skinny crescent to notice something interesting about the bright highlands regions of the Moon. Those regions have brighter patches and, here and there, long streaks of bright dust that radiate out from craters for great distances. A few of the Moon's less ancient craters have these *rays* extending out distances of hundreds of miles from them.

Moon Quadrant 4—the Southwest Quadrant. South is up and east is left, as in most telescopes.

The rays are best seen when the Moon is full or at a large phase, the time that the Sun shines as vertically down as possible on the lunar landscape. But one of the great initial recognitions by lunar observers is that the Moon near full is really too dazzlingly bright to view through a sizable telescope without a filter. Even *with* a filter or with a small telescope you notice that most of the features on the Moon are washed out and difficult to see when the Moon is near full. What is missing then are shadows in the landscape, shadows that help define the rims of craters and the verticality of mountains. Indeed, even when the Moon is a crescent, a telescope shows that detail is sharper near the terminator. In light of this fact, a good way to proceed with learning the features of the Moon in detail is to pursue what is visible near the terminator at intervals of every day, or every few days, during the lunar month. We will do this in our next chapter.

But first we need to discuss a lunar phenomenon that can be confusing to beginners yet is highly important: libration.

Not Quite the Same Face

Libration is an exception to a rule that even many complete beginners already know: the Moon always keeps the same face toward Earth. We are told this is because Earth's moon, like most moons in the solar system, is locked into *synchronous rotation.* This means that over vast periods of time Earth's pull has slowed down the Moon's rotation until that rotation became the same length as the Moon's revolution period. In other words, the amount of time it takes for the Moon to turn around once is the same as the amount of time it takes to complete an orbit around Earth. This guarantees that we always see the same side—the near side—of the Moon pointed at us.

But not *quite* the same side due to a phenomenon called libration.

Libration is a nodding of the Moon's face to left or right, also a little bit up and down. The latter—the *libration in latitude*—occurs because the Moon's rotation axis is not quite perpendicular to the plane of its orbit. As a result, we sometimes get to stare a little beyond the Moon's North Pole, other times a little beyond its South Pole. The slow nodding of the Moon's face to the left or the right is called the *libration in longitude.* It occurs because the Moon's orbit is not perfectly circular around Earth and the Moon therefore travels at different speeds during the course of a month. While the Moon travels at different speeds in different parts of its orbit, it always rotates at about the same speed. This means that when the Moon is traveling faster than average we can peek a little farther past its trailing edge. And it means that when the Moon is traveling slower than average we get to see a little farther behind its leading edge.

Thanks to libration, a lunar feature never looks quite the same two different times, even at the same phase each month. We are guaranteed endless variety.

Close-Up Views of Lunar Craters and Other Features of the Moon

Sight 36

In the previous chapter, we took an overall look at the Moon in low magnification. But another one of the best sights in astronomy is a close-up view of a particular lunar landscape. Instead of singling out one crater (I'd choose Copernicus) or one other kind of lunar feature (I'd choose the Straight Wall—maybe) or even one region (I'd choose the Mare Imbrium region), let's tour as many of the landscapes as possible.

While reading the following text, you can refer to the lunar maps distributed through the last chapter and this one. In these maps, south is up—the inverted view provided by most astronomical telescopes. But note that these maps and all the text in this chapter follow a convention of lunar observing that says that east on the Moon is the direction in which a person on the Moon would see the Sun rise. That seems to make sense. But confusion can arise because the official east—leftward on our maps—is the side of the Moon that is pointed west in our sky on Earth!

The Waxing Crescent Moon

If you need to refresh your memory about the Moon's phases, about crescent and gibbous and why they appear where and when they do, refer back to Sight 20. For explanations about the Moon's age, see Sight 21.

We start our tour with the waxing (growing) lunar crescent.

The first extremely prominent feature you may notice as the Moon grows past two days old (two days from New Moon) is the elliptical-looking Mare Crisium. It is the only mare on the Moon's near side that is completely surrounded by bright highlands. Actually, how roundish or skinny Crisium looks varies greatly during a month and, if viewed at the same phase, from month to month—thanks to libration (see the previous chapter). Crisium is so outstanding an indicator of libration that you can even judge the extent and direction of libration with the naked eye by looking at this "Sea of Crises."

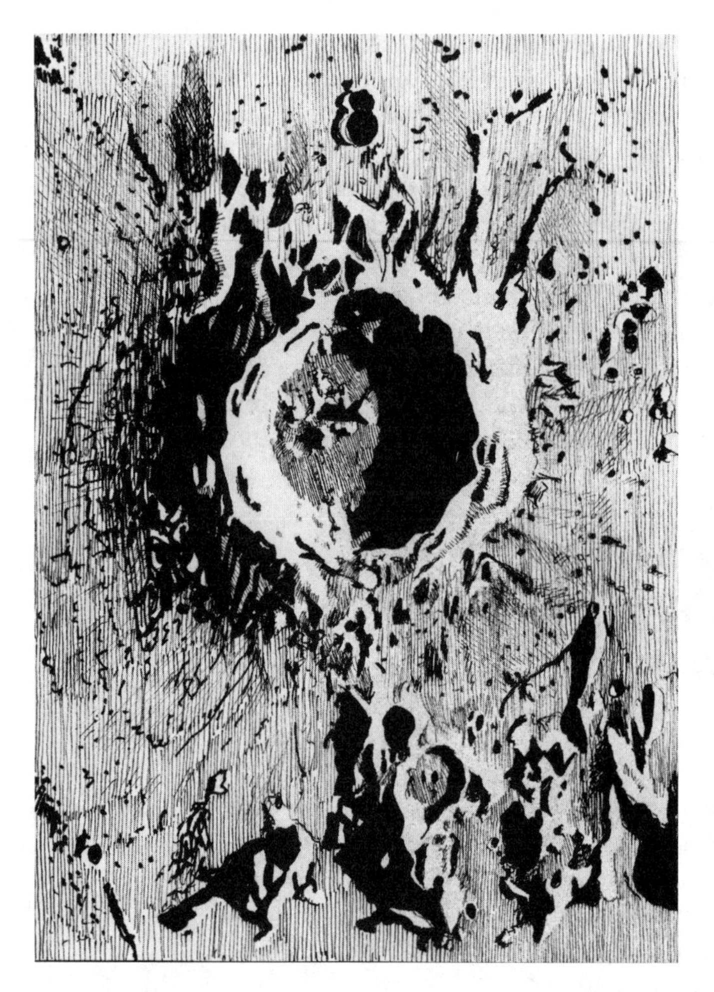

The lunar crater Copernicus.

Even through a finderscope you can glimpse a line of three great craters near the terminator well south of Mare Crisium when the Moon is about three days old: Langrenus, Vendelinis, and Petavius.

Around day four, the small but brilliant crater Proclus is visible just west of Mare Crisium and extends bright rays across the mare. Mare Fecunditatis is now fully in view and displaying a pair of little craters Messier and Messier A that will grab our attention because of the prominent double ray that will start glowing from the latter crater as the Sun gets higher over this part of the Moon. During the next day or so, we get a good view of Fracastorius—once a crater, now a bay on the edge of Mare Nectaris. Farther south you can see the crater Rheita and the Rheita Valley (the Rheita Valley is really a chain of craters 115 miles long and 15 miles wide). Toward the opposite (northern) end of the Moon, the superb pair of craters Atlas and Hercules are now visible.

By a lunar age of six days, not only all of Mare Nectaris but also all of Mare Tranquillitatis is lit. So is most of Mare Serenitatis. But of course the Sea of Tranquility is famous as the destination for the first landing of humans on the Moon. Many fine craters are now well placed for observation: Aristoteles and

Eudoxus; Plinius and Menelaus; and Maskelyne. But most impressive is Theophilus, with its 18,000-foot terraced walls and multiple central mountain mass. It intriguingly overlaps the older crater Cyrillus and with Cyrillus and Catharina it forms a rugged western edge to Mare Nectaris. Not far from this line of craters is the Altai Scarp.

The Moon reaches First Quarter and is half lit at about an age of seven days. The terminator now runs across many stunning craters and mountains. The Apennines mountain range is now partly in view. It is the Moon's most spectacular, running 600 miles with peaks up to 18,000 feet high. Within about the next day the southwestern end of the Apennines catches the sunrise and creates a projection of light along the terminator that even the naked eye can see. There is a break between the northeastern end of the Apennines and the start of the Caucasus Mountains. This break forms a strait between Mare Serenitatis and Mare Imbrium. The latter is still mostly in darkness, but light has spilled far enough into Imbrium to reveal the impressive pair of craters Aristillus and Autolycus. A little farther north, the Alps are showing, with a dramatic cut of darkness—the great 80-mile-long Alpine Valley—through them. The southern highlands are filled with numerous major craters and now, when the low Sun throws features into sharp relief, it is the time to identify two huge but eroded craters: Hipparchus and Albategnius.

From First Quarter to Ten Days Old

More spectacular lunar features come into view in the several days after First Quarter than at any other time of the month.

From about day seven through day nine, the terminator is crossing much of the most scenic sea on the Moon, the Mare Imbrium. The southern border of Imbrium is formed by the Apennines, whose entire length is now visible. Beyond the southwestern end of that range is the beautiful crater Erastosthenes and beyond that the most magnificent crater on the Moon: Copernicus. Copernicus has been called "the Monarch of the Moon." It is brilliant and possesses an amazingly complex interior, with three central mountain masses and an extensive area of landslide debris inside its walls, which are up to 17,000 feet high in places. A few days after the terminator passes it, Copernicus glows even brighter. It stands in contrast to the dark-floored Plato, which sits in a peninsula of highlands that separates Mare Imbrium from Mare Frigoris. On the floor of Mare Imbrium itself there are the solitary mountains Pico and Piton, and the line of peaks called the Straight Range. All these mountains are now spectacular as the sunlight first catches their tops above the still-dark mare floor and then illuminates the entire area, casting the sharp shadows of the mountains dozens of miles across the plain. Four prominent craters within Mare Imbrium are Archimedes, Aristillus, Autolycus, and Timocharis.

On day nine, the terminator has already passed Sinus Medii (Central Bay), the small mare area that marks the center of the Moon's face. It has also passed a pair of ancient craters that sit south of Sinus Medii near the start of the rugged southern highlands: Ptolemaeus and Alphonsus. Other prominent craters near these include Arzachel, Alpetragius, and Herschel. Farther south are some of the Moon's most famous craters: the "young" and beautifully formed Tycho (its rays not yet lit up) and the old, foreshortened but enormous Clavius. Most of Mare Nubium is now in sight, with the very prominent crater Bullialdus at its northwestern edge. Near Nubium's eastern edge is the amazing Straight Wall, a cliff (actually a fault) that now looks like a dark line but that looks spectacularly different when it reappears after vanishing for days around Full Moon.

A heavily cratered region of the Moon.

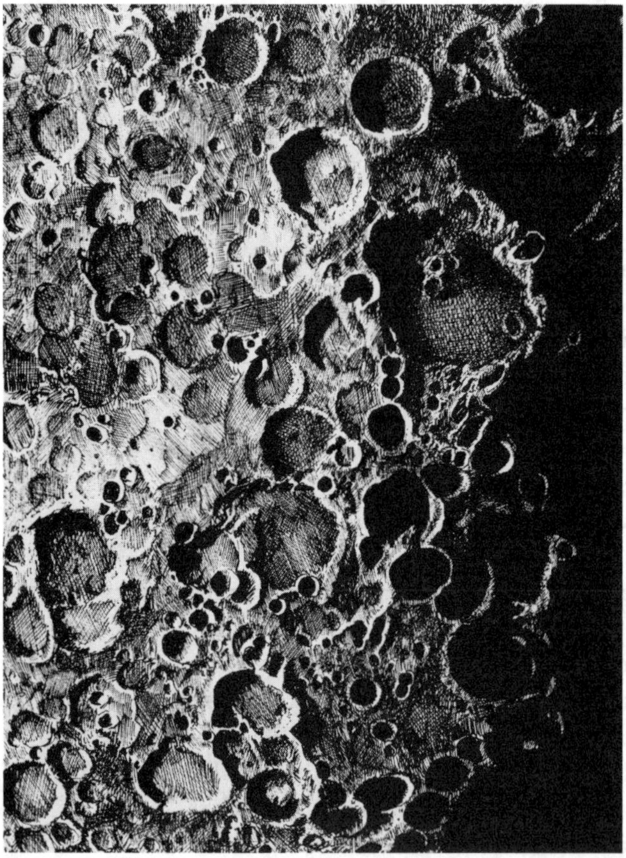

On day ten of the lunar month, the splendid Sinus Iridum (Bay of Rainbows) is at the terminator. This beautiful cove of Mare Imbrium may once have been a dark-floored crater like Plato, only considerably larger. On this day the bay's western border, the Jura Mountains, are hit by the rising Sun before the lower land and so gleam out like a "jeweled handle" in darkness. The full extent of the Carpathian Mountains, northwest of Copernicus, is now illuminated and forms a partial boundary between Mare Imbrium and the even more enormous but irregular-shaped "overflow" mare, Oceanus Procellarum (Ocean of Storms). For the past few days, the ray systems of the eastern part of the Moon, those of Proclus and little Stevinus (north of the smaller bright surround of the far larger crater Langrenus), have been prominent. But now the mighty ray system of Copernicus and the greatest ray system of all, that of Tycho, are beginning to grow in strength.

From Twelve Days Old through Full and After

On the twelfth day of the lunar month, Copernicus has been joined by two smaller but also brilliant craters in Oceanus Procellarum: Kepler and Artistarchus. Aristarchus has the greatest surface brightness of anywhere on the face of the Moon and both it and especially Kepler possess magnificent ray systems. Near bright Aristarchus, the crater Herodotus stands in dark contrast and from the northern wall of Herodotus runs for a long distance the winding Schroter's Valley. South of Kepler and Oceanus Procellarum, all the small Mare Humorum is visible, with the enormous crater Gassendi on its northern edge. In the southwestern region of the Moon, far past Mare Humorum, are the broken and worn craters Schickard and Schiller. South of Schiller, near the very edge of the Moon, and therefore extremely foreshortened (almost into being a line) is the ancient worn outline of the crater Bailly. It is difficult to identify even during a favorable libration but is worth the effort because this is the largest feature that is called a crater (or "walled plain") on the near side of the Moon. Bailly measures about 180 miles (290 kilometers) across.

At last we reach Full Moon, on the fourteenth day of the lunar month. Unless your telescope is small, a filter is necessary to reduce the strength of the light. There is little contrast now on the nearly shadowless Moon. The best thing to look at are the mighty ray systems, particularly that of Tycho. There is a ray that bisects Mare Serenitatis and it may be derived from Tycho—over a thousand miles away! Full Moon is also the time to get the best look at features near the western edge of the Moon. Here, you'll find Grimaldi, whose floor is the darkest on the Moon's face. Almost as dark is the inside of the crater Riccioli, which is even closer to the Moon's edge than Grimaldi.

After Full Moon, the sunset terminator slides across the Moon from lunar east to lunar west. The optimum visibility of most lunar features from the first half of the lunar month is now repeated—but the light of the low Sun is now hitting them from the opposite direction and there are important differences in appearance. One amazingly different appearance is presented by the Straight Wall. It appeared as a dark line after First Quarter then disappeared for several days around Full Moon. Now it reappears around Last Quarter as a breathtaking scratch of brilliance when the cliff's face, pointing to lunar west, catches full sunlight. As the Moon wanes, also notice the dimming of the previously bright ray systems. But one ray system that is at its best on the waning Moon is that of Byrgius (actually, of its little companion, Byrgius A), a crater located between Mare Humorum and the western edge of the Moon.

Sight 37

SUNSPOTS AND OTHER SOLAR FEATURES

In the past two chapters, we studied lunar features. But the Sun's surface sports a variety of "features"—or, rather, changing phenomena—that amateur astronomers can also enjoy. Of course, solar observation is not as easy as lunar observation. A direct look at the Sun, even with just the unaided eye, can cause serious damage to the vision. In this chapter, before we can describe sunspots and other solar features, we need to discuss the various techniques and equipment needed for safe observation of the Sun.

Solar Projection

Whereas it is safe to observe the Sun, even with a telescope, during the total stage of a solar eclipse, it is not safe to look at the Sun any other time, not even during a partial eclipse. As long as even a tiny percentage of the blindingly bright surface of the Sun remains visible, you must not look at it—even with just the naked eye (unless it is very dimmed by air and haze near the horizon and doesn't dazzle in the slightest). The word *blindingly* is meant quite literally. The eye itself does not feel pain when infrared radiation and especially ultraviolet light are causing burns to the retinas of the eyes. So pain won't warn you that damage is in the process of being done. Look at the Sun for even a second with the light-gathering power of a telescope and you are likely to experience either temporary or possibly even permanent blindness. Even the most beautiful celestial sight is not worth risking your vision for.

Naturally, there *are* ways that astronomers have developed for looking at the Sun safely.

The first way to observe the Sun safely is the one for you if you do not have solar filters. I refer to the technique of *solar projection.* The observer begins by putting his or her back to the Sun. A pair of binoculars or a telescope is then trained on the Sun with the observer making absolutely certain not to have his or her eye in the line of sight of the eyepiece or eyepieces. What you want to do is have an image of the Sun projected onto a screen of some kind—a piece of cardboard will do. How do you get the Sun into a telescope's field of view without looking into the telescope (which one mustn't do when the Sun is in it)? Just move the telescope until it casts the smallest possible shadow—a telescope tube will then produce a round shadow. When you achieve this positioning, the Sun should be in your telescope and its image projecting out onto your screen. You can vary the size and sharpness of your image by adjusting the focusing knob or by adjusting the distance of the screen from the telescope. (By the way, be careful when you are adjusting the distance of the screen—a telescope can bring enough heat to a focus at one spot to set the paper on fire!)

Did you know that you can form a very small image of the Sun to look at safely even without optical aid? All you need is one or two pieces of cardboard, with a pinhole pierced through one of them. Standing again with your back to the Sun, hold the cardboard with the pinhole so that it is perpendicular to the incoming sunlight. On some surface—it can be something as simple as a second piece of cardboard—intercept the light that is passing through the pinhole. Is it really an image of the Sun? Yes, the image is too small to allow you to see even large sunspots on it, but it is big enough to reveal the "bite" taken out of the Sun by a sizable solar eclipse. (If you don't even have cardboard, how about using trees? The chinks between the leaves of most trees will project sun images of various sizes and sharpness on the ground below. These images are dappled sunlight in the shade of trees—look closely and you'll notice such dappling is mostly caused by ellipse-shaped sun images. They are elliptical rather than round because the Sun is rarely overhead—never from midnorthern latitudes—and its light strikes the ground at less than a 90° angle.)

Solar Filters

Solar projection can be fun but does not provide images as sharp or detailed as you'd like unless you have a very precise set-up (for instance, a rigid screen mounted to your telescope so that the screen's surface is perfectly perpendicular to the sun image coming out of your eyepiece). If you have some money to spend, the solution is solar filters.

Unfortunately, many people who wish to see a solar eclipse get misinformation about what is and isn't a safe filter for solar viewing. Even for just solar viewing with the eyes (no binoculars or telescope) things such as sunglasses, unexposed photographic negatives, and other commonly suggested materials are unsafe to use. Don't make the dangerous mistake of thinking that because the visible light of the Sun is greatly reduced you are safe. It's the ultraviolet light that is the greatest threat, and even sunglasses that are rated for UV protection are not intended for staring at the Sun. Enough UV will get through to harm your eyes.

What kind of materials, then, do enable a person to view the Sun safely? For viewers using just their eyes, there are two choices that cost only a few dollars.

One choice is cardboard glasses with built-in patches made of aluminized Mylar. You can find ads for companies that sell these in any large astronomy magazine or on the Internet (they are often called eclipse glasses, for people are usually interested in using them to view a solar eclipse—even though they can also help you see very large sunspot groups when the Sun is active). Always check eclipse glasses to make sure the Mylar is not damaged: you can hold them up to a bright indoor light to confirm that there are no pinholes or tears in them. A second safe solar filter for viewing with eyes only (no binoculars or telescope) is shade number 14 welder's glass. A big enough piece to look through comfortably will cost only a couple of dollars and can be purchased at or ordered through a local welding supply company. The only shade of welder's glass darker than number 14 is number 15—but it's too dark to see the Sun well through that. On the other hand, shade number 13 is a bit too light—don't use it.

Both Mylar eclipse glasses and welder's glass turn the Sun an odd color (blue and green, respectively), but that doesn't interfere with your using them to watch the progress of a partial solar eclipse or very large sunspots. If, however, you want to see numerous sunspots in glorious detail, you need to use a telescope—and, with it, solar filters made especially for this purpose.

There are several major manufacturers of solar filters for telescopes (again, look in the ads of an astronomy magazine or on the Internet). What's most important is that they attach to the telescope between the Sun and the main lens or mirror. Many years ago it was common for manufacturers of cheap telescopes to include a solar filter that was screwed into your eyepiece. If you come across one, do not use it. The danger of it becoming overheated and cracking is not negligible.

In recent years, there have been some amazing improvements of commercially available solar filters. There has even been marketing of telescopes—and surprisingly low-cost binoculars—that have built-in filters and are intended for solar observation only. Of course, not every beginner will want to spend considerable money for specialized viewing with these personal solar telescopes. Also expensive—though coming down in price—are *hydrogen-alpha filters.* These give amazing views of intricate detail on the solar surface and of prominences outside of the magical time of totality.

Sunspots and Faculae

Even with just solar projection it is possible to observe a few of the most important solar features and phenomena. Certainly sunspots—sometimes large numbers of them. But you can also notice that the edge of the solar disk seems dimmer than the center. This is not an optical illusion. You are seeing *limb darkening*. It occurs because near the edge of the Sun there is a longer pathway of gas for light to pass through—and therefore a greater amount of gas to dim the light. The darkening of the Sun's brilliant surface near the Sun's limb makes it possible to sometimes see the brighter areas called *faculae*. Faculae are bright clouds of hydrogen floating above the *photosphere* (the visible surface of the Sun). They are often associated with sunspots. By the way, do not confuse faculae with *solar flares*, brief bright outbursts that are rarely visible without special filters—unless you are very lucky and happen to catch a brilliant white-light flare.

Of course, the easiest and richest of solar features for amateurs to observe are sunspots. *Sunspots* are seemingly dark splotches on the photosphere that are really the surface manifestation of a cutting off of energy by a lower-down magnetic disruption. They are dimmer and cooler than the surrounding surface but only relatively so. Their temperature is still a few thousand degrees Fahrenheit, and if we could move a sunspot out to space by itself (and magically maintain its nature) it would appear as a blindingly bright object.

Everything about sunspots is changing, complex, and fascinating to watch. They may appear singly or in groups. Each is usually associated with another (though not always visible) spot (or spots); the two spots are thought to be

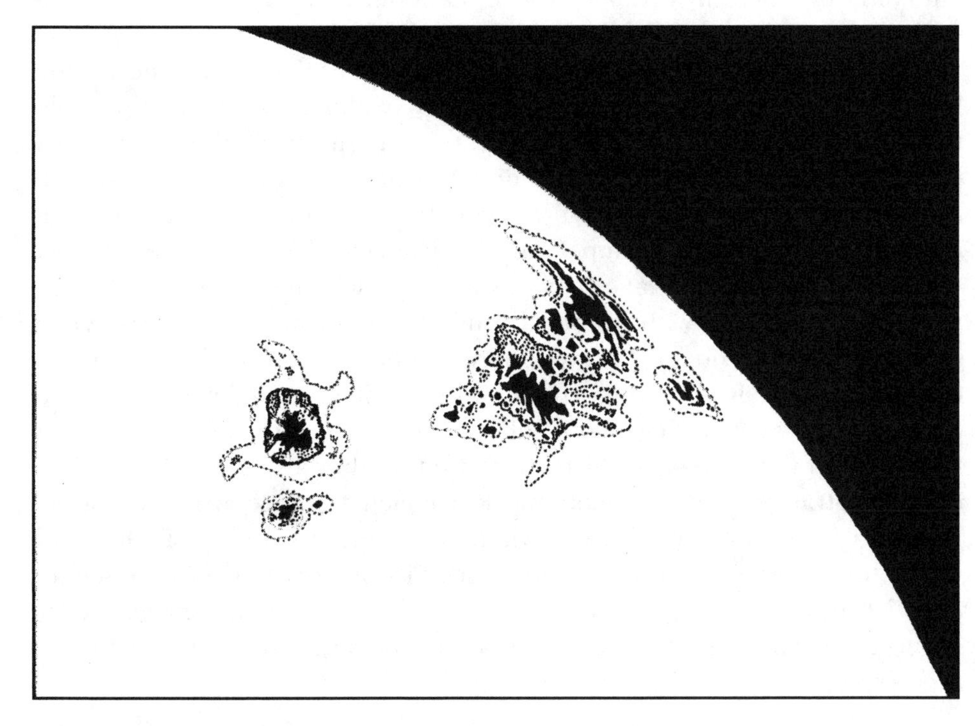

A view of sunspots near the Sun's limb.

magnetic exit and reentry points. Rarely can one spot be identified as being paired simply with another, however. Usually, it is a group that exhibits some signs of this symmetry in its pattern. If this is the case, it is known as a *bipolar sunspot group.* A spot may have a darker central area, called an *umbra,* and a lighter area around it, called a *penumbra* (these two terms are better known in amateur astronomy in their application to the central and peripheral parts of Earth's shadow during a lunar eclipse or to the Moon's umbra during a total eclipse of the Sun).

An entertaining activity for amateur astronomers who observe the Sun is counting the number of sunspots. The number of sunspots increases (with continuing short-term variations) as the Sun approaches the maximum in its eleven-year *solar cycle* of solar activity. During the cycle, the locations on the Sun where spots appear changes, too. At solar minimum, sunspots start appearing at middle latitudes on the Sun (though seldom farther north or south than 40° from the Sun's equator). Then, as the cycle moves toward solar maximum, the spots keep appearing at lower latitudes on the Sun (though rarely closer than 5° from the solar equator). Sometimes there are still a few sunspots from the old cycle at low latitudes when the first spots of the new cycle start appearing at middle latitudes.

If the Sun has poles and an equator, then it must have rotation. However, because the surface is not solid, the Sun rotates at different speeds at different latitudes. The Sun's equator rotates at about 26.7 days; at 40° solar latitude, about 29 days; at 60° solar latitude, about 31 days.

What is the average amount of time from a spot's first appearance to the fullest development of its group? About ten days. So it is sometimes possible to watch much or all of a sunspot group's evolution during the course of its passage across the near side of the Sun. Only occasionally is a region on the Sun disturbed enough to stay active throughout its trip around the far (hidden) side of the Sun and reappear on the near side (face).

Each day's first look at the Sun's surface, whether by projection or through filter, carries with it a wonderful element of surprise. Sunspots can sometimes change noticeably even in the course of an hour. So when a new day dawns, there is no telling what new developments you will see. As you get your first view, perhaps you will have a feeling that sunspots are not really part of the Sun—it's easy to fantasize that they are like flocks of birds or other strange objects (a celestial Rorschach test?) seen in silhouette. That is a beguiling thought, but when you remind yourself of what and where these features really are, the wonder is even greater. You are seeing 3,000° F hot patches made to look black. You are seeing a magnetically induced "storm" of evolving gloom that may be larger than our Earth—or even Jupiter.

PARTIAL ECLIPSES OF THE SUN

Sight 38

Every total eclipse of the Sun or the Moon includes a partial stage. But unfortunately every partial eclipse does not continue on to the climactic sights of solar or lunar totality. When we call an eclipse partial, we usually mean that is the largest extent of coverage that occurs: just part of the Sun or Moon's disk is hidden or dimmed. The Moon can even experience an eclipse that is milder (lighter) than partial—a penumbral eclipse (see Sight 18).

We are left with the undeniable conclusion that partial eclipses are runner-ups in any contest for most exciting sights. A partial lunar eclipse falls considerably far short of the beauty and impressiveness of a total lunar eclipse. A partial solar eclipse falls tremendously short of the shattering, awesome beauty and strangeness of a total solar eclipse. Having admitted the limitations of partial eclipses, however, we should hasten to point out that they are still among the best sights in astronomy. When a total eclipse is not available at your home site—or perhaps anywhere in the world—but a partial eclipse is, you will want to see the partial eclipse.

Despite mentioning partial lunar eclipses here, I have elected to include my detailed discussion of them in connection with total lunar eclipses—the subject of Sight 18. But we are left here with the wonders of partial eclipses of the Sun.

Safety First, Then the Wonder

Whereas it is safe to observe the Sun, even with a telescope, during the total stage of a solar eclipse, it is not safe to look at the Sun during a partial eclipse. As long as even a tiny percentage of the blindingly bright surface of the Sun remains visible, you must not look at it—even with just the naked eye (unless it is dimmed by air and haze right on the horizon and doesn't dazzle in the slightest). For all the information you need about safe techniques such as solar projection and safe products such as solar filters, refer to the previous chapter, where you will even find some specific advice about using devices as simple as pieces of cardboard or the trees to project views of a partial solar eclipse.

But now, trusting that you, good reader, will heed all the words of precaution about solar observing, I will proceed straight on to the wonder of a partial solar eclipse.

A sequence of views at the 1984 annular-total solar eclipse.

If there is something amazing about being able to summon a detailed image of the Sun for safe viewing from out of the blinding solar blaze above, there is even greater magic in tuning in to a "live" drama involving that Sun: the majestic crossing of the Moon in silhouette across its face.

The first "bite" out of the edge of the Sun caused by the Moon is almost immediately detectable. I recall rushing outside as clouds broke just in time to permit me a quick look at the solar eclipse of April 8, 2005. At my location, the Moon was due to never reach more than 3.6 percent of the way across the Sun—would the dark indentation be visible on the low Sun by projection with a small telescope? I hastily set up, got the Sun's image onto a piece of cardboard, and immediately saw—with a thrill of recognition—the small but quite prominent bite. I even used eclipse viewing glasses and through the aluminized Mylar, with no magnification, I was able to detect the slightest out-of-roundness of the Sun's disk in just the right place.

Even after you've observed a number of partial solar eclipses previously in your life, there is always that moment when you first detect the dark indentation of the Moon and feel that surge of satisfaction and excitement. The immense interworkings of Sun, Moon, and Earth have functioned perfectly again, and the Moon has arrived right on time for its rendezvous. Is the Moon due to creep a third of the way across the Sun at this eclipse from your location? Halfway or more? A large part—but far from all—of the daytime hemisphere of Earth is observing some version of this event. There are usually sunspots for you to watch the Moon's edge creep up on and perhaps to hide—a drama within a drama.

The time of maximum coverage comes and you note it—wishing wistfully that it was more, dreamily that it was 100 percent and the miracle of a total eclipse was in store. But regaining your composure, you continue tracking the Moon's trek and feeling fulfillment in having been able to witness this stately—yet also eerie—celestial performance. You know that the next such show may be years away, or even longer, if you are clouded out on those future days when the Moon will again meet the Sun.

Annular and Hybrid Solar Eclipses

What is the best that partial solar eclipses get? Very, very good indeed. Admittedly, at most of them you will see no more than what I just described. The eclipse will probably be one of your better or even one of your best astronomical sights of the year, but probably not one that will loom large in the astronomical memories of your life (unless it was your first solar eclipse). But there is a special kind of partial eclipse that could be very high on your life list: an annular eclipse of the Sun.

An *annular eclipse* occurs when the Moon passes centrally across the face of the Sun but is a little too far out in its orbit to cover the entire blinding face of the Sun. Consequently, there is a blazing *annulus*—that is, a ring—of light left visible. We can still consider this a form of partial eclipse, albeit a special one. It's even possible for the Moon to be so far out in its orbit that the ring is thick, and roughly 3 percent of the Sun's diameter is left uncovered around all sides of the Moon. But that is still enough coverage to cause changes in the landscape and the sky like those that occur not too many minutes before totality. Did I say the ring in such an annular eclipse is thick? It doesn't seem thick when you are staring at it through eclipse glasses or projected onto a screen.

I know because I did both at the distant, 94 percent magnitude (94 percent of the Sun's diameter covered) annular eclipse of May 10, 1994. My wife and I drove to Ithaca, New York, for that one and observed it from the green pastures of Cornell University. We set up a small telescope at a quiet spot on campus and were ignored until just before the time of annularity. Glitters in the landscape had turned to mere gleams; the sky had darkened somewhat and turned an odd color. Suddenly, a student stopped and asked if she could watch the eclipse with us. She was from India and had once seen a total solar eclipse. She volunteered to hold the cardboard screen I was projecting the solar image on. In the last minutes before annularity, a small crowd of students gathered around us. The horns of the thin crescent solar image started to extend farther and farther—a bit more, a bit more, and then, like magic the cusps joined and became a ring. I also found it thrilling to look up with the eclipse glasses and see what I knew was a ring of fire in the sky. The few minutes of ring eclipse passed so fast; the ring broke on its opposite side, the cusps retracted. The hush lifted and the several dozen people around us murmured—some of the murmurs were thanks to us—and then quickly dispersed to their classes and dorms, chatting happily with one another. We thanked our friend from India and she thanked us, and then we went our separate ways—a fond relationship and a memory for all life, all based on maybe twenty minutes centered on a ring of fire in the sky.

Now imagine that you are at an annular eclipse where 99.8 percent of the Sun's diameter is covered. You certainly get to see some corona, maybe Baily's Beads, and, amazingly, not a diamond ring (see Sight 2) but a diamond necklace—a wire-thin ring of fire with blazing "stars" strung here and there around it. This can happen at a very large annular eclipse or at one that is classified as an *annular total eclipse*: total along part of its path but annular at either end (because the surface of the Earth there curves away from the Moon enough for the tip of the cone of the Moon shadow to just barely fail to reach the surface). The figure on page 196 shows a sequence of images taken at the annular eclipse of May 30, 1984, when 99.8 percent of the Sun's diameter was covered. Note the diamond necklace—or was it "just" a necklace of Baily's Beads?—in the last shot.

Annular and partial eclipses can be spectacular in a different way if they occur when the Sun is low. First, because the low Sun gets flattened due to refraction, and this can lead to some bizarre distortions of the annular ring or crescent Sun. Second, because you may be able to look at the eclipse directly with the unaided and unprotected eye if it is on the horizon and dimmed enough by horizon haze.

It's also possible to glimpse the shape of the eclipsed Sun directly and safely with the unaided eye if the Sun is shining through a thick enough deck of clouds. One has to be *very* careful about making such an observation, though. There is a temptation to stare too long, and clouds may be thinner than one thinks—and very permeable to ultraviolet radiation. But on October 3, 1986, I used my judgment and looked—briefly. I was feeling disappointed because it appeared as though the partial solar eclipse that day would be completely clouded out. I'd almost given up. But then, suddenly, my living room brightened with (still somewhat veiled) sunlight. I rushed to the door to try to catch an observation. And up there, through clouds still thick enough to be protective, I saw, for just a few seconds before overcast hid all, something astonishing. Without a constraining and discoloring filter or secondhand imaging by projection, I saw the real, natural Sun in the real natural sky with, an alteration of its normal appearance. A big top part of the Sun was missing, cleanly and sharply sliced off.

The whole experience was so sudden and thrilling that both the experience and the appearance of the Sun itself suggested to me something I had read about poetry. The great poet Emily Dickinson once wrote that the way she could tell that something was real poetry was if it made her feel like the top of her head had been knocked off. Seeing the Sun with the top of *its* head seemingly knocked off made me feel that way, too. Apparently, this eclipse was poetry.

TRANSITS OF MERCURY AND VENUS

Sight 39

A transit is the passage of one astronomical object in front of a much larger one. Quite a few times a year, amateur astronomers with medium-sized telescopes can see transits of the various moons of Jupiter and the shadows of the moons of Jupiter across the face of the giant planet (see Sight 41). Much rarer, however, are transits of the only two planets that can go directly between Earth and the Sun and therefore be viewed as a dark dot on the solar face (with proper protection of the eyes, as described in Sight 37).

One of the planets in question is Mercury, which transits the Sun as seen from a large area of Earth about eleven times a century. But far more rare—and spectacular—is the passage of the other planet, a planet that appears about five times wider than Mercury and is encircled by a heavy atmosphere (Mercury has almost none at all). The planet is Venus, and its transits usually occur in pairs, with the second event taking place eight years after the first. The problem for would-be observers is that the time between pairs is more than a century!

Fortunately, we currently live in the eight-year period between two events in a Venus transit-pair. If you are reading this book before June 6, 2012, you can still hope to see a transit of Venus in your lifetime. If you're reading this book soon after that date, there will still be Mercury for you to see. But you're going to have to live a mighty long life to catch the Venus transits in 2117 and 2125!

Transits of Mercury

There is no other celestial event that is really like a transit across the Sun. Sunspots, if they survive long enough, take well over a week to be carried around the Sun and out of our view. A transit across the Sun lasts merely hours and features the planet transiting dramatically past sunspots, or even in front of some of them. Sunspots are sometimes roundish, sometimes irregular in shape. Transiting planets are virtually perfectly round and opaque dots. Last but not least, sunspots are temporary disruptions of the solar surface. The black dot we see crossing the Sun in a transit is an entire world, one of two unique and enduring (for billions of years) planets crossing some part of the gulf of space between us and the Sun.

Let's start with the transits of Mercury. Settling for little Mercury instead of Venus is not a heart-rending disappointment. It is in fact what every observer of a transit in the twentieth century had no choice but to do (the last Venus transits before those in 2004 and 2012 were the pair in 1874 and 1882).

Mercury is very small against the face of the Sun. But at least we're seeing it bigger than we do when it is at greatest elongation or other commonly observable positionings. Even if you observe a transit of Mercury with Mercury and the Sun low in the sky and seeing is far from ideal, you can still get a passably good perception that this 12"-wide spot is too round and dark to be any sunspot. And movement of Mercury relative to the Sun—and relative to any sunspots that *are* visible—quickly gives it away.

Clouds have deprived me of two transits of Mercury that would otherwise have been visible from my geographic location. But so special are the transits I did see that I remember the details of all of them very vividly. The transits of Mercury I have viewed were the ones on May 9, 1970; November 10, 1973; and November 15, 1999. In the vast era of our history that we are now in, transits of Mercury can occur only within a few days of these dates in May and November.

Perhaps the most unusual of my three transits of Mercury was the one in 1999. It was due to begin not long before sunset, and clouds were entirely hiding the Sun. A friend and I were at my dear old East Point site, looking west across Delaware Bay. There was a break in the clouds farther down near the horizon. Would it endure until the Sun got down there and Mercury was entering the edge of the solar disk? Yes, it would. And we were ready to watch it by both projection and through a solar filter on an 8-inch telescope. The seeing was better than we would have guessed so low in the sky and with clouds near our line of sight. Mercury was visible as a strange little bite on the Sun's limb; then, suddenly, it was wholly on. Clouds came and went. It looked as though the horizon itself was now too cloudy for us to follow the event that low. But, one more time, there was some clearing down there. The suspense was terrific. The edges of the Sun began to boil from turbulence, and Mercury's shape became a little distorted. But my last glimpse of the dark spot of the planet came when part of the Sun was already below the horizon. Amazing—I would never have thought the world's little dot would have remained visible until almost the final possible moment.

Transits of Venus

As this book is being written, the next transit of Venus—the last for 105 years—looms ahead on June 6, 2012. Its early stages will be visible at day's end from where I live in the eastern United States. It will in this way be a perfect complement to the one that occurred on June 8, 2004, and whose end was visible here early in the morning.

All my life I had been waiting for the 2004 transit. Some veteran skywatchers had paid to travel to Europe to get the transit high in their sky. I actually didn't mind having it occur only at sunrise and for an hour or so after—even though I knew that seeing might be bad with the Sun low in the sky. But the question was: would the big morning be clear here? And if not, would it be possible to take off in the car the night before to drive to where dawn *would* be clear? At first, weather prospects were looking pretty bad. I considered driving a few hundred miles north or south. But the prospects in those locations didn't sound good, either. Was I going to miss this first of the pair of transits, maybe both of them?

The forecast improved greatly the day before and I committed to observe it not at the seashore, where I was afraid there might be fog, but from a surprising location: the seventh-floor balcony of my mother-in-law's apartment in a nearby small city. My wife and I showed up early. There were a few light clouds, but mostly the haze was going to prevent me from seeing the Sun right on the horizon. But that was true pretty much all up and down the eastern seaboard. There was still a chance, however, that a few minutes after sunrise the haze would permit what I most longed to see: a naked-eye view of the dot of Venus on the Sun's face. I was convinced that Venus would be large enough to see without magnification in front of the solar disk. But not all experts agreed with that assessment. Who would prove right?

A television suddenly showed a live image from a local news team's helicopter that was a few dozen miles to the east—and there was the Sun, coming into view out of the haze. Sure enough, within a minute or two the view was being reenacted out on our balcony: we were getting our first glimpse of the red disk of the Sun ever so slowly burning bright enough to become visible. What followed was perhaps as much as 5

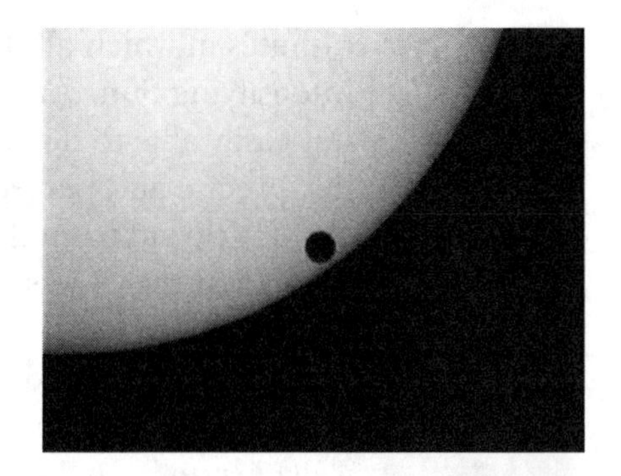

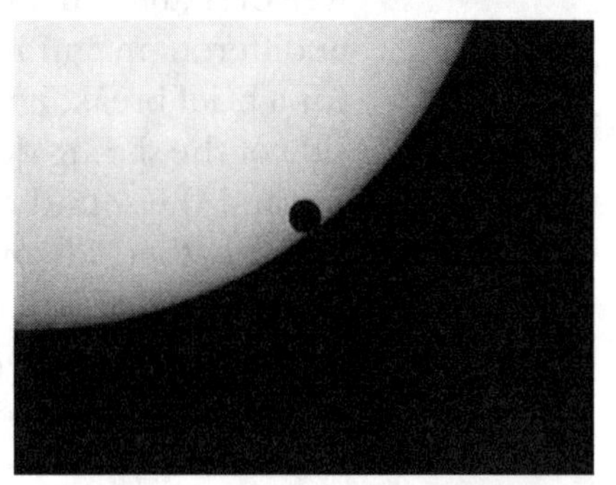

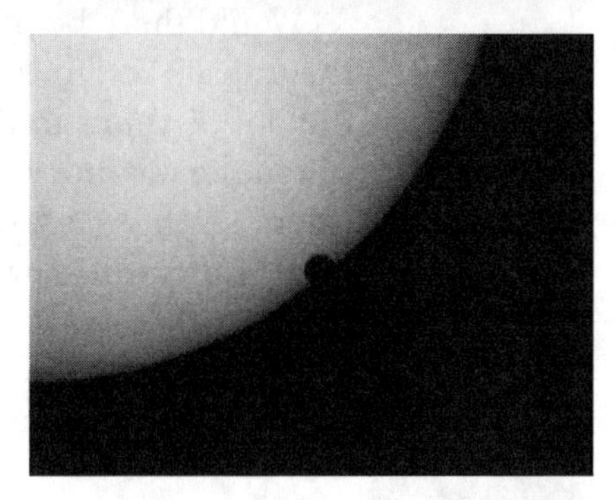

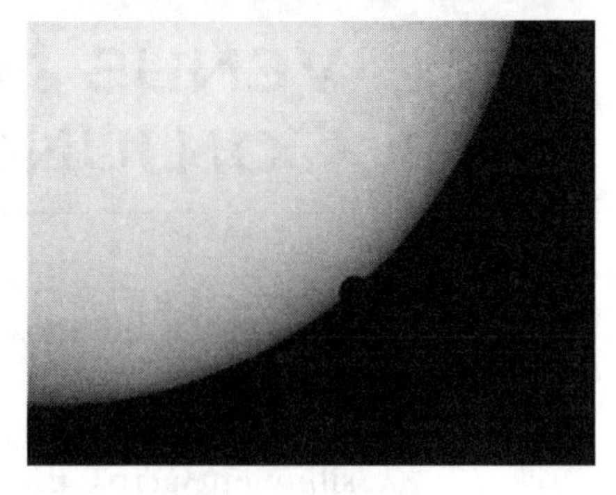

A sequence showing the egress of Venus during its 2004 transit.

minutes in which all three of us could safely gaze at the steadily clarifying and intensifying Sun. And even my mother-in-law, who was sixty-nine at the time, was easily able to detect the longed-for sight with her unaided eye: the incredibly precise and perfect dot of Venus on the disk of the Sun! Observing without optical aid from that balcony intensified for me an awesome gut reaction. I truly felt that I was perched a bit precariously on one planet and seeing—almost feeling across the gulf of space—the definitively solid body of another planet hanging between me and the Sun.

After the Sun got too bright, it was time to turn to solar projection. For well over an hour I spent almost every minute watching that big planetary dot at different magnifications. The few times I stepped back into the apartment for a brief break, it was a glorious exercise in unbelievable contrasts. On one side of the sliding door, a cozy living room and kitchen, with the smell of coffee and the sound of conversation going on in comfy everyday domesticity. On the other side of the door, a monumental heavenly lineup of three celestial bodies that had last occurred 122 years earlier. No person alive on Earth had seen a transit of Venus before this day. After this day, I may have been one of 100,000 who saw it in the sky or directed onto a screen, and one of millions who viewed it on television.

Would there be the historically reported "black drop effect" at the end of the transit? Yes, for me at least, a bit of a seeming waist of black did connect the dot of Venus to the edge of the Sun and then thicken until the planet's silhouette was an ever-dwindling bite out of the Sun's edge. In a lot of places where seeing was excellent and a precise telescope-filter combination was used, observers saw no black drop at all. But many people—including myself—were stunned at transit's end by a sight we really hadn't expected: when the dot of Venus was partway off the Sun, the delicate arc of illuminated Venus atmosphere still completing the circular form of the planet!

Venus Near Inferior Conjunction

It is by far the most thrilling passage that any planet frequently makes: the flight of Venus from the evening sky, past the Sun, and into the morning sky. Of course, if the transition includes a transit—a journey of Venus in silhouette across the face of the Sun—that is a whole additional level of fascination and awe. But, as we saw in the previous chapter, transits of Venus are spectacularly rare.

Crescent Venus near a church steeple far away.

That's all right, however. The sights of Venus leading up to and/or away from inferior conjunction are always dramatic and beautiful. The fall from the evening sky can be breathtaking, and the vault into the dawn sky equally marvelous. If you see the planet in the two different venues (dusk and dawn) on the same day, how unique, precious, and soaring a perception. If you catch the view of Venus as a long narrowest dazzling arc in the telescope, it is thrillingly lovely—as is the crescent being just resolved with binoculars. And if you can manage to see the shape of the crescent Venus with your naked eye—the form of another world in the sky with unaided vision? You will feel you are the privileged recipient of a magic vision revealed to few.

The Telescopic View

Back in Sight 11, we discussed Venus at greatest elongation, and the figure on page 62 showed why the planet displays different phases as it reaches various points in its orbit relative to Earth. You may wish to take another look at that diagram now. What we need to understand for this chapter is that as Venus races along the curve of the near side of its orbit, we see its disk get ever closer and therefore bigger—but also we see more and more of its nighttime side. In other words, the hemisphere of Venus facing us becomes more the night side of the planet, with less of the planet that is illuminated by sunlight remaining visible. When Venus passes very near the Sun in the sky, not only is it lost in the solar glare but also its sunlit crescent may become too thin to see—unless of course it passes right in front of the Sun with its night side entirely facing us and we see its silhouette as a transit.

Bearing in mind the orbital diagrams and positions, let's follow what we observe of Venus as it goes from greatest evening elongation through inferior conjunction to greatest morning elongation.

At greatest evening elongation, we see a half-lit Venus about 24" across in the telescope. At greatest evening brilliancy, we see a quarter Venus (about 25 percent illuminated) about 43" across in the telescope. In the evening sky about two weeks before inferior conjunction, we see in the telescope a roughly one-twentieth Venus (that is, Venus about 5 percent illuminated) about 58" across—or should we say 58" long, for it is then a very skinny crescent. Inferior conjunction occurs, but Venus is usually unviewable then. After inferior conjunction the previous stages occur, in reversed order, before sunrise: Venus in the morning sky two weeks after inferior conjunction; greatest morning brilliancy; and greatest morning elongation.

What are special sights to look for in the telescope at each of these stages? Back in Sight 11, we noted that it is fascinating to try to determine the exact night that Venus appears precisely half lit. We could now add that for a while before, during, and after greatest elongation it is a good time to try to glimpse very subtle shadings in the clouds of Venus. As Venus dwindles to a much thinner crescent, try looking for a strange general or patchy faint glow on the dark part of the planet—if you see some, it could be the mysterious (still unexplained) "Ashen Light." As Venus becomes an extremely thin crescent, look for deformations and dark spots at the ends of the crescent. These may usually just be optical effects—for instance, caused by poor seeing in our own atmosphere. But it's just possible that one of these cusp-caps or other odd appearances of one of the crescent's cusps (points) really is a glimpse of the reduced cloud cover near the poles of Venus. Of course, you don't have to identify a rare and possibly illusory phenomenon like a cusp-cap to be thrilled by the sight of a breathtakingly thin arc of Venus. That arc has much greater surface brightness than the very thin crescent Moon. And though it is 30 times smaller than the Moon, it is still the largest—or, rather, longest—sight of a planet we ever get to see—it looks huge at a high magnification.

What is the closest to inferior conjunction that Venus can be seen in a telescope? Right at the moment of inferior conjunction (those few who have pulled off this feat have generally done it, with great care to avoid getting the Sun in their eyepiece, by viewing the planet when it is high in the daytime sky near the Sun). What is the thinnest crescent Venus that can be seen in a telescope? Not counting observations during a transit, it is possible to see Venus when it is much less than 1 percent illuminated. In fact, if Venus is close enough to the Sun, observers will see the horns of the crescent extended around until they meet—a ring of light made possible by sunlight shining through the atmosphere of Venus.

Thrilling Binoculars and Naked-Eye Views

All the previous observations should be tried with Venus as high as possible in the sky. The higher the planet, the better the "seeing." But they should also be attempted as soon after sunset (or before sunrise) as possible—or even in the daytime sky. The reason is that the contrast between the surface brightness of Venus and a night sky—even a midtwilight sky—is so great that the planet dazzles the eye, making any chance of seeing cloud features impossible and even perceiving a sharp edge to the crescent itself difficult.

This advice to see Venus high and in a bright sky is equally (or even more) important to heed if you are trying to detect the crescent as a tiny object in binoculars or as a truly minute form with your unaided eye. But first let's talk about the beauty of just viewing Venus as a point of light with binoculars or naked eye when it is approaching or receding from its (usually) unviewable inferior conjunction.

The amount of time between greatest evening elongation and greatest brilliancy happens to be virtually the same as the amount from greatest brilliancy to inferior conjunction: about five weeks. How many hours or minutes Venus sets after the Sun and how high it is in the sky at each stage of twilight—both of these depend on more than just where Venus is in its orbit and how many degrees it appears separated from the Sun. As we saw back in Sight 11, the visibility of Venus also depends—very strongly for observers at middle latitudes—on the time of year and the resultant angle of the zodiac with respect to the horizon. When Venus nears inferior conjunction in late summer or early fall, its angular separation from the Sun is slanted at a very shallow angle to the horizon. Its visibility is poor because even if it is still 30° from the Sun (compared to the maximum of about 47° that occurs at greatest elongation), that 30° is mostly to the side of where the Sun sets, placing the planet low in the sky even as early as sunset. On the other hand, if inferior conjunction is going to occur in early spring, the angles are excellent—almost all the elongation is vertical.

The opposite times of year from those just listed are favorable and unfavorable for seeing Venus soon after inferior conjunction, in the east before sunrise. In other words, when inferior conjunction occurs in late summer, the planet will need only a matter of days to start rising enough minutes before the Sun to be seen, almost directly above where the sunrise will occur. But a dawn elongation of Venus is shallow and low in late winter.

Let's suppose we have a favorable time of year for Venus nearing inferior conjunction. What will its behavior be like for the naked-eye observer at mid-northern latitudes? Venus makes a stunningly rapid departure from the evening sky. In the best-case scenario, such as what happened in 2001 and recurs at eight-year intervals (thus next in 2009), we find that one month before inferior conjunction Venus is still setting more than 3 hours after the Sun and is still more than 36° high at sunset. Three weeks later (about ten days before inferior conjunction), Venus has become only half as bright and sets just 1¼ hours after the Sun, appearing only 15° high at sunset. Four days later (only about six days before inferior conjunction), Venus still sets 40 minutes after the Sun and is only about 7° high—but is *still* easily visible to the naked eye if the sky is quite clear. What's interesting is that the northerliness of Venus at this showing makes it by this date already conceivably visible before dawn too—technically, the planet is still east of the Sun, but it now rises about 20 minutes before sunrise (though binoculars may be necessary for glimpsing it). During the remaining days before inferior conjunction, observers can attempt the feat of seeing Venus with the naked eye (or binoculars) both before sunrise and after sunset *on the same day*. Clouds foiled my last attempt to do this back in 2001, but I did something perhaps as exciting:

I saw Venus with the naked eye after sunset one day and before sunrise the next. Venus thus appeared to me at *both* the gateway into night and the gateway out of it.

Now, we return at last to the question of trying to observe the actual shape of Venus in binoculars or with the naked eye. The former is not too difficult. You can achieve it a few weeks before Venus reaches inferior conjunction, possibly even with very steadily held binoculars (I've been successful by simply propping myself against a tree)—though preferably with mounted binoculars. The planet only needs to be about 40" or 50" wide for you to succeed—as long as you observe it high and under good seeing conditions.

But what about seeing the crescent of Venus with the unaided eye? That may be a feat that can only just barely be done by people with extremely sharp vision—some of them state they are really only seeing that Venus (a small point of light in a still-bright sky) is elongated in the correct direction. But in 2001 I used a special trick to succeed in getting a view of the tiny crescent without using magnification. I punched a hole only about 1 millimeter wide in a piece of cardboard and centered it precisely in front of my eye. The peripheral regions of the lenses of our eyes have slight distortions (apparently caused mostly by the pull of muscles that attach them to the eye). These are what make stars and planets appear to have "rays." By limiting my vision to just the center of my eye's lens, I was able to get the sharpest possible view of Venus and see the minute crescent exquisitely! (It's too bad that this trick also greatly reduces the light getting through to one's retina—a problem for seeing objects a lot dimmer than the brilliant Venus.)

Jupiter and Its Moons

Sight 41

Jupiter is the most dependable of planets in the sky and, almost always, the most dependably intricate and rich in the telescope. It is the second brightest planet, after Venus. But unlike Venus it can be visible all night long, not just a few hours. Another advantage Jupiter has over Venus is that it does not get lost to view for months at a time because of proximity with the Sun—only for about two weeks, once a year. Jupiter is also dependable in the sense that it takes about twelve years to complete one orbit around the Sun and around our heavens. That means Jupiter basically spends one year in each of the twelve classic constellations of the zodiac—what could be more orderly and memorable?

For about half of every year, the naked eye can see yellow-white Jupiter rise, brilliant and steady shining during the hours of the night. Only occasionally—briefly every few years—can one other planet, Mars, do the same with a brightness and color that rivals the appearance of Jupiter. But it is in its telescopic appearance that Jupiter really excels. In your eyepiece, its globe presents a greater illuminated area than those of all the other planets combined, and more details than all the other planets combined (save for Mars on the rare occasions when it comes quite close to Earth). Add to this the attraction of Jupiter's four big and brilliant moons and their endless variety of positionings and you have a planetary banquet more sumptuous by far than any other.

First Views of Jupiter in the Telescope

First-time observers are often disappointed by the size of planets in a telescope. They should not be with Jupiter—especially if they move up from the low-power view to a much higher magnification image. Of course, the size of your telescope's aperture and the steadiness of a given night's atmosphere place limits on how high a magnification you can use before images get fuzzy and no more details are seen. That said, on a good night Jupiter can loom really large in a high-power eyepiece of an 8-inch or even 6-inch telescope. Seeing this, an observer can well believe that he or she is looking at a world eleven times wider than Earth and so massive that it contains more material than all the other planets put together. But making Jupiter look as big as you can is a mere gimmick. The real thrill is what you can see on the planet, and your introduction to that should come at fairly low magnification.

At a rather low power (say 45× or 90×), Jupiter is such a bright piece of palpable celestial fruit it deserves the term *globe* or, even better, *orb*. You may notice that this orb is not perfectly spherical, though: it is flattened at the poles and wider at the equator. And it should be easy to tell what is Jupiter's equator. For a medium-sized (or smaller) telescope will almost always show at least a few grayish bands striping across the planet. You naturally think that such bands are horizontal and somehow associated with the planet spinning around in the direction of their orientation. And you are correct. You are seeing a patterning of clouds that are being stretched out at different latitudes into globe-encircling strips. As a matter of fact, Jupiter is not only the largest planet but also the one that rotates the fastest. The length of a "day" (equivalent to the 24-hour period on Earth) on Jupiter is less than 10 hours.

The darkish bands on Jupiter are known as *belts*; the light bands between them are known as *zones*. But before we closely examine them and other features in Jupiter's clouds, there is another Jovian sight (sight related to Jupiter) that is already drawing your attention—maybe even before the fact that Jupiter has belts. That sight is the four great Jovian moons—the Galilean satellites of Jupiter.

The Dance of Jupiter's Moons

These four moons, working from innermost to outermost, are Io, Europa, Ganymede, and Callisto. When you first view them in a small telescope, you will be reliving the experience of their great first recorder and chronicler: Galileo. Like him, you will see one, two, three, or four of the four "Galilean satellites" at once, arranged in some beautiful pattern more or less in the equatorial plane of the planet. Even an hour or two reveals changes in their arrangement—sometimes all are on the same side, sometimes there are two to a side. Sometimes two of the moons come very close together (in the sky—one will always be many thousands of miles beyond the other in space). They look like intense little stars, though as you increase your magnification they start looking a little fat and then, with a 6-inch or larger telescope in good seeing at 200× or 300× power, even like tiny dots. But the first question that presents itself when a beginner sees less than the promised four moons is, of course, where the others are. The answer is: either in front of the planet (but lost in its brightness), or behind the planet, or in the shadow of the planet.

These phenomena of the Galilean satellites are explained in the figure on page 210. There are actually four major kinds of events to explain. A *transit* of a Jovian moon occurs when the moon passes in front of Jupiter's disk. A medium-sized telescope may or may not show the little moon during a transit, but to see such a sight is thrilling and mesmerizing. Actually, an event that is usually easier to see is a *shadow transit*—the passage of one of the moons' dark shadow across the contrastingly bright face of Jupiter. On the other hand, if one of the moons enters into Jupiter's shadow, the event is an **eclipse**. Finally, if a moon passes behind Jupiter, we call the event an **occultation**. By the way, when Jupiter is near opposition (opposite from the Sun in the sky), we don't see eclipses of the Galilean satellites because they coincide with occultations—in other words, when a moon enters Jupiter's shadow it is also behind Jupiter from our point of view. The figure on page 210 shows the more interesting situation that occurs when Earth is not almost directly between the Sun and Jupiter and therefore does not have the same line of sight to Jupiter as the Sun does. In this case, Jupiter's shadow is projected a bit to the side from our point of view, and a satellite can disappear into and/or reappear out of the shadow before it disappears behind or emerges from behind the planet.

Incidentally, for one period of quite a few months every six years (half of

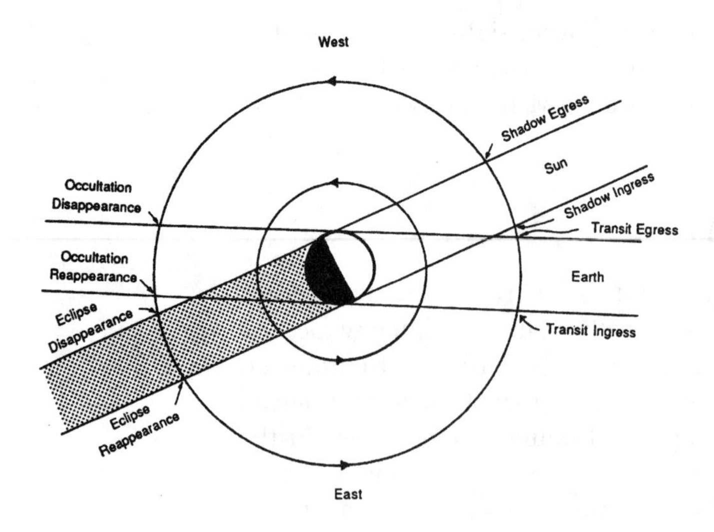

Jupiter's Galilean satellite phenomena. The globe is Jupiter. The zones marked "Sun" and "Earth" are the lines of sight from the Sun and Earth, respectively. These lines of sight often coincide more closely than shown in the particular situation of the diagram. When they coincide closely, eclipses occur during occultations and therefore cannot be observed.

a twelve-year Jupiter orbit) we get an extra treat: the Galilean satellites eclipsing and occulting one another. Many of these events are difficult to observe but well worth trying.

Even when the Galilean satellites are not eclipsed or eclipsing, transiting, or shadow-transiting, they are always enthralling. All of them are technically bright enough to see with the naked eye but are usually overwhelmed by the glare of their parent planet. Few people have seen Galilean satellites with the naked eye. But the outer two, Ganymede and Callisto, can definitely be discerned without optical aid when they are near their greatest elongations (maximum angular separations) from the planet. I myself have glimpsed both Ganymede and Callisto with my naked eye.

In the past few decades, space missions have taught us so much about these four huge moons. Io and Europa are similar in size to Earth's Moon but the gravity of the nearby giant Jupiter has done amazing things to them. Io is the most volcanically active world known, its eruptions so frequent that scientists believe it has literally turned itself inside out during the course of its existence. Europa's flat topography is icy, but the ice shows odd cracks and it is believed that an ocean of water—and just possibly life—exists some distance below the surface. Meanwhile, Ganymede is the biggest and Callisto the third biggest moon in the solar system. More than 3,200 miles wide, Ganymede is bigger than the planet Mercury (though much less massive), and its surface displays somewhat Earthlike tectonics. Callisto is spectacularly rich in cratering, and its reddish color is faintly noticeable in fairly large amateur telescopes.

Where do you get information about where the Galilean satellites will be tonight and which phenomena of them—eclipses, occultations, transits, or shadow transits—will occur? Such information can be read from the "corkscrew diagram" of the planet not only in certain annual publications as *Astronomical Calendar* (see the sources section) but also in monthly publications—the major astronomical magazines.

Festoons, Garlands, Rifts, Ovals, and the Great Red Spot

On a good, steady night, an 8-inch or larger telescope can show many belts and zones on Jupiter—though rarely as many as shown in the figure below. Don't expect to be able to detect most of these on a typical night. Bide your time; learn to see better by using sketching as a tool to train your eye. In the meantime, can you see any traces of color in the belts (they sometimes show some ruddiness) or the zones? More important, can you see any of the features other than belts and zones that exist in Jupiter's clouds?

Among the most important are *festoons* and *garlands*. They are subtle projections from out of the dark belts into the light zones. A festoon totally bridges a zone. A garland doesn't but may form a hook or even curve all the way back to form a loop. The opposite of these features is *rifts*. A rift is a bright bridge of cloud that entirely crosses a belt from one zone to another. A *knot* is a thickening or darkening in a part of a belt.

Rather different kinds of features in Jupiter's atmosphere are *ovals* and *spots*. Ovals are noncircular and are often very bright and fairly large. Spots are rounder and more sharp-edged but relatively small.

There is, of course, one feature on Jupiter that is the most famous of all, and it is not small: the Great Red Spot (GRS). This feature is an eye-shaped "storm" that is much larger than Earth and has lasted for at least centuries. It is located at about 20° S on Jupiter, which is just south of the usually very prominent South Equatorial Belt (SEB).

Unfortunately, beginning telescope users often expect too much from this legendary whirl. The GRS is seldom a very prominent feature, and its color is often quite pale. Over the decades, the

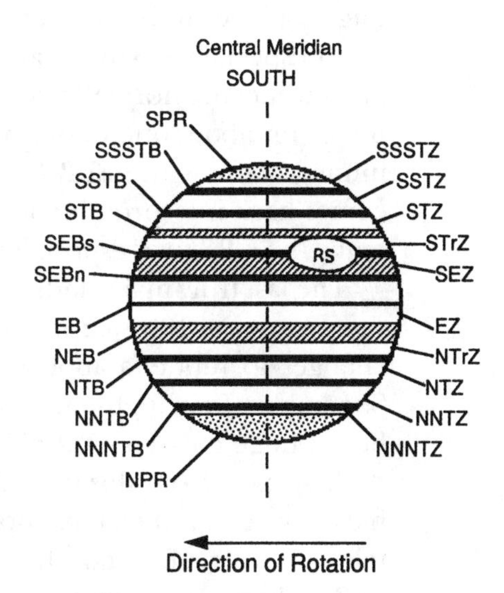

Jupiter's belts and zones.

INDEX TO ABBREVIATIONS

ZONES

EZ - Equatorial Zone
SEZ - S. Equatorial Zone
STrZ - S.Tropical Zone
STZ - S.Temperate Zone
SSTZ - S.S. Temp. Zone
SSSTZ - S.S.S.Temp. Zone
NTrZ - N. Tropical Zone
NTZ - N. Temperate Zone
NNTZ - N.N. Temperate Zone
NNNTZ - N.N.N. Temp. Zone

BELTS

EB - Equatorial Belt
SEBn -S. Equatorial Belt North
SEBs -S. Equatorial Belt South
STB - S. Temperate Belt
SSTB - S.S. Temperate Belt
SSSTB-S.S.S. Temperate Belt
NEB - N. Equatorial Belt
NTB - N. Temperate Belt
NNTB - N.N. Temperate Belt
NNNTB - N.N.N. Temp. Belt

OTHER

SPR - S. Polar Region
NPR - N. Polar Region
RS - Great Red Spot

GRS does have exciting periods when it becomes more prominent and colorful, although even then the "red" is more like a salmon pink or other mild hue. When the GRS is pale, you may instead glimpse the gap it sits in: the Red Spot Hollow. What's important from night to night is knowing when the GRS will be passing the central meridian on the face of Jupiter. That is when we get a straight-on view of it or any other Jovian cloud feature. Because Jupiter rotates so fast, looking even two hours before or two hours after will give us a much more foreshortened and difficult view of the GRS. Where do you find the times of GRS crossings of the central meridian for each night? Again, the astronomical magazines have the information. You should note, however, that such predictions are approximate when given a few months in advance because the GRS does drift in longitude, and at varying speeds!

The changes in Jupiter's cloud patterns are ceaseless. The face of the mighty planet never looks the same. In fact, over the years there are times when especially dramatic alterations to its appearance are seen. Once in a great while, the normally most prominent belt—the SEB—will lose its prominence for about one to one and a half years. When it does, the GRS—which impinges upon the SEB—gets more visible and colorful. When the SEB comes back into prominence, it typically does so boisterously and seems to make the GRS turn especially faded and nearly colorless.

The last truly major fading of the SEB occurred back in 1989–1990. But in the summer of 1994 a tremendously rarer event made spectacular temporary changes to Jupiter's appearance. That was when the fragments of Comet Shoemaker-Levy 9 plummeted into the Jovian clouds. The result was a number of huge dark spots at different longitudes but at the same latitude. The darkest spots from this probably once-in-a-millennium event were easily visible in even very small telescopes and stayed fairly prominent in medium-sized telescopes for a few months.

The heavens never cease to offer surprises. One further Jovian example came in February 2006. That was when an amateur astronomer imaged Jupiter and was the first to discover that a cloud feature that had grown from the merging of several spots had suddenly turned the same color as the GRS. This feature, dubbed Red Spot Jr., is only half the size of the GRS. But that's still a little larger than Earth. Red Jr. survived a close encounter with the GRS in the summer of 2006 but there will be more encounters, and we don't know what will happen. We do know that this is the first large spot ever seen to turn the color of the GRS.

SATURN AND ITS RINGS AND MOONS

Sight 42

When people get their first proper look through a good astronomical telescope, there are two sights that never fail to make them literally gasp or yell out. One is the surface of the Moon. The other is Saturn.

Everybody knows that Saturn is the planet with something truly extra: a spectacular set of rings. Astronomy writers have frequently called Saturn "the showpiece of the solar system." So anyone preparing for their first view of Saturn through the telescope has very high expectations. The amazing thing is that those expectations are almost always exceeded.

The Precise, Perfect, Artistic, 3-D, and Live Rings

Even in a small telescope, the ball and the rings of Saturn are so perfect, so exquisitely precise in their embodiment of a vast artistic geometry, that the sight staggers belief. When I had just turned thirteen years old, I got my first really decent telescope and trained it on Saturn. The plastic dust cover on one end of the telescope was engraved with an attractive little picture of Saturn. When my brother and I saw the real Saturn, we joked that it was a similar little picture already in the telescope—because the perfection of the real Saturn seemed almost unbelievable.

The planet Saturn.

All first-time (and often even thousandth-time) observers of Saturn are struck by this feeling of wonderful disbelief. There are a number of elements that go into making Saturn in the telescope even more thrilling than Saturn in the best photographs. There is the three-dimensionality sensed when one sees not a picture on a flat piece of paper but a ball floating within a ring floating within a sea of darkness. This is especially powerful when we can see that the shadow of Saturn's globe is cast upon the rings. There is also the knowledge that you are viewing this live and direct. You are staring not at a photograph but at the masterpiece itself.

The Nature and Tilt of the Rings

When confronted with a ragingly beautiful and mysterious sight, it is only natural to find yourself wondering what it really is, what the explanation is for the strange vision confronting you.

The first telescopic observers had instruments so poor they couldn't even identify what the form of the rings was. Galileo and others could distinguish merely bizarre projections—"ears"—sticking out from either side of the planet. And then the projections grew harder and harder to see and disappeared for a year or two—only to return and grow in prominence. So to the

Saturn's rings at different tilt angles.

mystery of what their form was, there was the added mystery of why they vanished for a fairly lengthy period every so many years.

The first person who built himself a telescope excellent enough to perceive the true shape of the rings was the mid-seventeenth-century astronomer and all-around genius Christiaan Huygens. He saw clearly enough to realize that the rings nowhere touched the planet, but rather encircled it with their glory. What was their composition? Even today astronomers debate the particulars of this question, but it was long ago realized that the rings could not exist as continuous sheets of material. In fact, the rings are composed of countless individual particles, at least in part icy, that revolve around the planet at different speeds according to their distance from it. The size of the particles ranges from ones as tiny as those in cigarette smoke up to boulders and mountains and even moonlets. Are these particles the leftovers of moons that couldn't form or of ones that spiraled too close in and were crushed by Saturn's gravity? The elements of this question are still being debated.

But what about the mystery of the disappearing rings? That mystery was quickly solved. Better telescopes showed that the rings are more or less visible depending on their angle of tilt toward Earth. The tilt of the rings remains the same in relation to Saturn itself—the rings are in the equatorial plane of the planet. But the rotating planet and with it, the rings, are somewhat tilted with respect to the plane of Saturn's orbit. Depending on where Saturn is in its orbit, the rings can appear to us tilted at a maximum of 27° for a while, but the tilt then decreases over the course of 13.75 years, reaches a minimum angle, then increases for 15.75 years before reaching maximum tilt again. Add 13.75 and 15.75 together and you get the orbital period of Saturn—about 29.5 years (the shorter duration—13.75 years—occurs in the part of Saturn's orbit closer to the Sun, when the planet travels somewhat faster). Notice I haven't stated what the minimum value for the tilt of the rings is. In case you don't know and haven't guessed: the minimum tilt is 0°. We see the northern face of Saturn's rings during part of its orbit and the southern face during the other part. But that figure of 0° means that the rings become edge-on or, as it is more often expressed, "edgewise." And the appearance of the rings of Saturn then is truly astonishing.

We could understand if the rings were thin enough when seen edgewise for them to become invisible in the earliest inferior telescopes. But the rings are so thin that they even may pass from view, or almost so, in today's largest and best telescopes.

How can this be? The span of the rings that we easily see from Earth measures over 170,000 miles in two dimensions. But the rings measure only a few hundred yards in the third. The span of the rings is almost twice as great as the diameter of Jupiter. It would fill most of the space between Earth and the

Moon, utterly dwarfing them. But when viewed from the side these rings are no thicker than our most towering skyscrapers on Earth are tall.

Basic Observations of the Rings and the Globe

During the vast majority of Saturn's three-decade-long orbit, you can see several beautiful features of the rings even with a small telescope. But let's assume you have a 6-inch or 8-inch telescope. If so, you will want to move up in magnification—preferably to well above 100×, if conditions permit. At the start of this chapter I wrote that Saturn in a telescope almost always exceeds a first-time observer's expectations. But there's one thing about the view that does sometimes disappoint the beginner: the small apparent size of the planet and the rings. Saturn is almost a billion miles away from us. It's almost twice as far as Jupiter. The rings are still vast enough in span to usually extend across an apparent diameter as great as, or greater than, that of Jupiter. But the globe is never more than 20" wide—compared to an average of about 40" and a maximum of 50" for Jupiter. So low magnification shows Saturn and rings only as perfect but minute objects. Use a little more power.

When you do, you should notice several things about the rings. First, that they *are* rings—not just a ring (singular). Even if they initially appear to be one object, note that the inner, wider part is much whiter and brighter than the outer part. This inner part is a separate entity, the *B ring* (think B for *b*roader and *b*righter). The outer ring is the *A ring*. The line of darkness—(almost) empty space—between the B and A rings is visible in good seeing in telescopes as small as about 4 inches when the rings are generously tilted and in larger telescopes even when the rings are pretty far from maximum tilt. The black line is called *Cassini's Division*. When you detect it, you are beholding a gap only about as wide as the Atlantic Ocean from almost a billion miles away.

Another ring sight is pretty easily visible a few months before and a few months after Saturn is at opposition: the shadow of the planet on the rings (around opposition, the shadow of Saturn falls directly behind it from our point of view, so we don't see the shadow). The addition of the shadow makes the planet and the rings seem even more three dimensional than usual.

There's more to see of the rings and in them under very good conditions or with slightly larger (10-inch or 12-inch) telescopes. But what can a medium-sized telescope show us of the planet itself?

The rings can be distracting for observations of Saturn's globe. This is especially true of the shape of that globe. Saturn is the most *oblate*—squashed at the poles, bulging at the equator—of all the planets. But the rings are spectacular and elongated in the equatorial plane of Saturn. As such, they can make it hard to notice the globe's oblateness—it is more obvious when the rings are near edgewise. Why is Saturn the most oblate planet? It spins almost as rapidly as Jupiter but is even less dense overall than that largest of the gas giant planets. The average density of Saturn is less than that of water, leading to the amusing suggestion that Saturn would float in a bathtub if you could find a tub large enough for it.

Another interesting aspect of Saturn is its color: a pale green-gold in smaller telescopes and a sort of butterscotch in medium-sized instruments. Visible against this color is, usually, at least one gray belt near the planet's equator. Saturn's *belts* (dark) and *zones* (bright) are less numerous and conspicuous than those of Jupiter. This is not only because the planet is so much farther away from us than Jupiter but also because it is farther away from the Sun and its atmosphere is less stirred by the Sun's heat. Still, a number of belts and zones may be seen on good nights (especially with a larger telescope)—and you may also notice the shadow of the rings (usually a rather thin line) on the planet. Don't forget to study whichever pole of Saturn is best displayed. The poles are covered with dusky caps that vary in darkness over the course of months and years and sometimes may show a bluish tinge.

Suppose you have a 10-inch or larger telescope and one of those rare nights when Earth's atmosphere is almost calm all the way through above you. What further wonders of Saturn's rings can you see?

Even though spacecraft visits have revealed that there are at least seven major rings of Saturn, most are too subtle to see with Earth-based telescopes. The exception besides the obvious A and B rings is the *C ring*, or *Crepe ring*. Its intensity seems to vary over the years, but it can often be glimpsed even in medium-sized telescopes if atmospheric conditions are excellent. The C ring appears as a sort of gauzy region of radiance between the B ring and the planet.

There are also divisions other than Cassini's that can be glimpsed at rather high magnification in superb seeing. These gaps tend to vary in intensity under the influence of Saturn's moons (including at least one that is *within* a gap in the rings). The second most famous gap in the rings is much narrower than Cassini's—it is *Encke's Division*, located out in the A ring.

Finally, there are strange sights of Saturn's rings around the time of edgewiseness. Matters are wonderfully complicated by the fact that Earth passes through the ring plane at a different time than the Sun does. So it is possible for a while for us to be looking at the dark side of the rings. Furthermore, there is usually not just one passage of Earth through Saturn's ring plane but three. The series of three occurs because Earth's motion in its orbit—which is slightly tilted with respect to Saturn's and the rings—is quick enough to let us get south of, then north, then south of the ring plane (or south-north-south) during the course of nine months. Unfortunately, the next two times Saturn comes to the proper place in its orbit—in 2009 and 2025—the timing will be such that we get just one passage of Earth through the ring plane. Even so, marvels will abound for telescopic viewers. I've seen for myself that with a good 6-inch telescope and excellent seeing that it is possible to glimpse the sunlit rings less than a day from edgewiseness. They appear thinner than the thinnest razor blade—in fact, like two spikes that almost look like individual rays of a star sticking out to either side of Saturn. For weeks or more around the edgewise presentation, we can sometimes see thicker areas of the rings showing up as clumps or beads of light on the otherwise smooth, straight, shining line of the rings. We must make sure not to confuse these interesting clumps with beads or specks of light that shuttle back and forth along the rings. These specks or stars are moons of Saturn orbiting back and forth past the planet in essentially the same plane as the rings.

The Moons of Saturn

When the rings are edgewise, the overall brightness of the naked eye point of light that is Saturn is approximately cut in half. Furthermore, the elimination of the rings' bright wideness permits the observation of Saturn's inner moons that are too dim for amateur astronomers to see at other times. However, we don't have to wait until the next edgewise presentation in 2009 to glimpse an impressive number of Saturnian moons.

How many moons can we see? All save one of Saturn's moons are smaller than all of Jupiter's big four Galilean satellites. But Saturn has *six* moons that are visible in 6-inch and 8-inch telescopes. And with larger telescopes, two more moons can be seen.

The figure below shows the orbits of the brighter Saturnian satellites approximately as they are oriented to us a few years away from a ring-plane

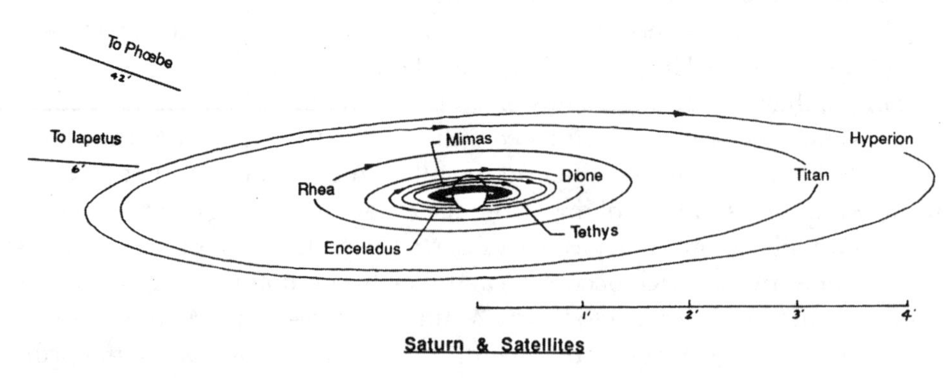

The orbits of observable Saturnian satellites. The apparent shapes of the orbits open and close during the course of the Saturnian year.

passage (edgewiseness of the rings)—as they are, for instance, in 2007. Imagine these ellipses becoming thinner and thinner as we near a ring-plane passage.

For current positions of these moons, consult the corkscrew diagram in popular astronomical magazines or a few good annual publications such as *Astronomical Calendar* (see the sources section in back of this book).

Eight Saturnian moons are bright enough to see every year (even when the rings are not near edgewise). Iapetus has a wide range of brightness—its brightness depends on which side of Saturn it is on. This is because on one side of the orbit the Sun catches the hemisphere of this moon that is bright as ice and on the other side of the orbit the Sun shines on the hemisphere of Iapetus that is black as coal. These and other mysteries of the Saturnian moons are starting to be answered by the images and other data being obtained by the *Cassini* spacecraft now orbiting Saturn. The spacecraft's most famous act came early in 2005 when the *Huygens* probe it had released successfully landed on the surface of arguably the most interesting moon in the solar system: Titan.

Titan is bright enough (8th magnitude) to be seen in binoculars, but a medium-sized telescope will begin to show its orange color. This moon, second only to Jupiter's Ganymede in size, is the only moon to have a truly massive atmosphere. And Titan continues to be considered one of the few worlds in our solar system that just might possibly harbor native life.

Mars at Closest in a Telescope

Sight 43

Mars—the very name sends chills of wonder down the spine of anyone who cares the least bit about outer space, and with good reason—or, rather, many reasons.

The reasons visible to the naked eye we detailed back in Sight 12. There we discussed how Mars got very much brighter every other year when it would reach opposition to the Sun in our sky. But we also noted that the tiger-colored planet kept getting brighter at each of these oppositions until, once every fifteen or seventeen years, it reached an extraordinary peak in brilliance. This occurred at the perihelic opposition—the opposition taking place when Mars was near the perihelion (closest to the Sun in space) point of its lopsided orbit.

The planet Mars.

Those awesome strange brightenings of a planet whose color suggested blood were regarded with fear in ancient times, and this planet was associated with the god of war in many cultures (the god of war in Roman myth was Mars). But starting in the seventeenth century when telescopic observers began getting their first look at the globe of Mars, the planet inspired fascination and fear for a new reason. As we discuss the details of observing Mars with a telescope when it is close, we'll examine these modern reasons why Mars burns more intensely in the human mind than any other world beyond our own.

One word of warning—or rather joyful challenge—is called for. Mars is the planet whose telescopic features ask the most of an observer. If you don't rise to the challenge of studying Mars carefully, and maybe even training your eye by using your hand to sketch its features as you glimpse them, you will see only its most conspicuous features—and those only on very favorable occasions. On the other hand, if you let Mars teach you to become a skilled planetary observer, you will see literally dozens of fascinating and changing details on its surface and in its atmosphere. You will become a true explorer of a planet beyond our own—without leaving your backyard.

The Planet Most Like Earth

When you look at Mars in your telescope, you will often find it disappointing. For many months at a time it appears as a tiny orange dot or speck, without the slightest detail. Its apparent diameter compared to that of other bright planets when they are farthest is only about one-fourth the width of Venus and Mercury, one-fifth the width of Saturn, and one-tenth the width of Jupiter. But just wait until Earth starts catching up to this slower, outside neighbor of ours. Mars can then grow almost a hundred times brighter and up to 7 or even 8 times wider in our telescopes. Even at distant oppositions where it grows to only about 5 times its minimum apparent size, we cross a threshold of visibility for surface features that makes all the difference.

All the difference? This featureless speck is revealed to be an orange globe rotating in almost the same amount of time as Earth, with almost the same amount of axis tilt as Earth, sporting polar ice caps that look dramatically like those of Earth and one thing else: seemingly green patches that grow more prominent when spring comes to the Martian hemisphere they are in. This is what astronomers started seeing of Mars as telescopes improved in the eighteenth and nineteenth centuries. They could see for themselves that Mars was the planet most like our own and suspected that the green areas were large regions of vegetation that were invigorated by water running down from the melting ice cap in spring.

In the nineteenth and early twentieth centuries, scientists came to realize that conditions for life anything like we know it were borderline at best on Mars. That is not only because Mars is much farther from the Sun than Earth and is therefore very cold, but also because Mars is only half the size of Earth and its gravity can support only a very thin atmosphere. But a thicker atmosphere in past eras of the planet could have retained more heat. So speculations began that life could have evolved on Mars long ago and then found ways to adapt to the deepening cold and thinning air.

Such speculations were on the mind of the American millionaire socialite-cum-astronomer Percival Lowell when he began to think he was seeing long thin lines connecting the green areas. The Italian astronomer Giovanni Schiaparelli had seen such lines (most of them later proved to be a kind of optical illusion) and called them *canali*—a word that should have been translated into English as "channels." Instead, it looked like the English word "canals"—artificial waterways. Lowell fell prey to the optical illusion of seeing these features when Mars came close and believed that they were indeed canals—canals built by an ancient, still-surviving race of Martians to conserve and direct the planet's dwindling water supplies. What followed Lowell's publicity of this scenario was H. G. Wells's book *War of the Worlds* (dramatized on radio so believably in 1938 by Orson Welles that it set off a panic that Martians really were invading New Jersey). In the science-fiction stories and movies of the first two-thirds of the twentieth century, by far the most common origin for imagined extraterrestrial invasion was Mars. The saying "little green men from Mars" became well known.

Our unmanned spacecraft visits to Mars in the past forty years have shown us just how harsh conditions on the planet are. The Martian atmosphere was shown to be about 95 percent carbon dioxide because several things that converted that gas to other substances in the history of Earth didn't happen on Mars. One that didn't occur, apparently, was the development of plant life that could change much of the carbon dioxide into oxygen. Most—though not all—of the ice in the polar caps and frost detected by the *Viking* spacecraft was "dry ice"—that is, frozen carbon dioxide rather than frozen water. A problem for life not fully appreciated before spacecraft visited Mars was the planet's lack of an ozone layer and resultant high doses of sterilizing ultraviolet radiation. But it's certainly not time to give up on the idea of Martian life. Discoveries made in the past few years have reopened the possibility that life once existed, just maybe even still exists, on Mars.

This long—and excitingly continuing—history of humankind's search for life on Mars percolates through us when the Red Planet comes close. We begin to glimpse many of its surface features in our backyard telescopes and perceive for ourselves the ways in which Mars resembles Earth.

The Surface and Atmospheric Features of Mars

What are the green patches on Mars that in some cases seem to darken with the coming of spring and lighten with the arrival of autumn? First, they're not really green. When an observer perceives this shade, it is really an illusion produced by contrast with the surrounding orange, leading our eyes to produce a complementary color. Call these areas instead the dark regions of Mars. And what they are is simply regions of darker rock and soil than the majority of the Martian surface, which is orange with fine sand. The changes in the extent of the dark areas—the so-called wave of darkening seen as a polar ice cap melts in spring—is really a wave of brightening of the orange areas that makes the dark regions appear darker by improved contrast. What causes the brightening of the orange areas is a melting of frost that reveals brighter, fresher deposits of the dust that were moved by the previous year's dust storms.

In any case, it is also true that the exact shape of the dark regions changes from year to year and a map like that of the figure below is only approximate. You should consult the astronomical magazines and their respective Web sites for which longitudes of Mars will be pointed at you during the hours when the planet is high for observation. The fact that the Martian day is only 37.5 minutes longer than Earth's means that a given face of Mars is presented to the observer about that much later on each succeeding night.

Which are the most prominent of the dark regions on Mars? Most prominent of all is the roughly triangular dark patch known as Syrtis Major. It was first noticed by Christiaan Huygens in the mid-seventeenth century. But other dark areas may almost rival it in prominence, especially at certain oppositions. An example of a dark region that varies a lot and in some years is far more vivid than others is Solis Lacus, "the Lake of the Sun." Solis Lacus stands out so well at some oppositions that it has earned the nickname "the Eye of Mars"—for its form can indeed look like an eye.

Most of Mars appears orange or, perhaps more accurately, ochre in the telescope. Some of these bright areas are more conspicuous than others. For instance, roughly south of the dark equatorial Syrtis Major is a large, bright, circular patch called Hellas. Spacecraft explorations show Hellas to be a 1,000-mile-wide impact scar, its rim highly eroded but its depth still as great as 3 miles. When winter comes to the southern hemisphere of Mars, a bright coating of fresh frost can form in Hellas, making it especially prominent.

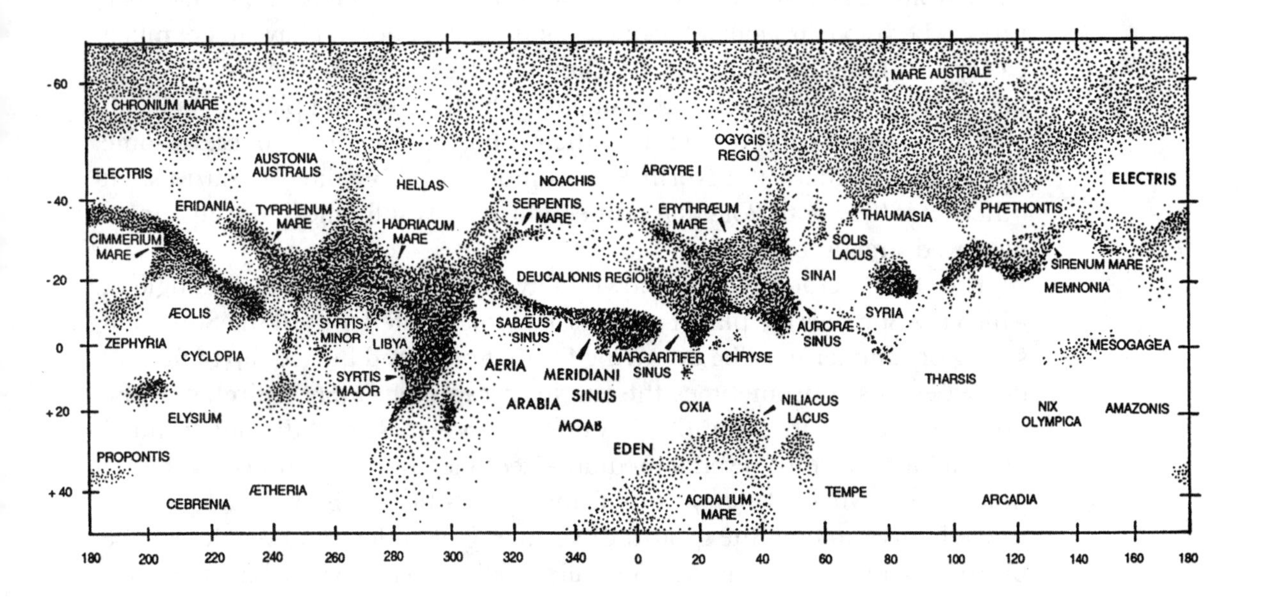

Telescopic surface features on Mars.

Mars is the only planet whose surface features are easily detectable in telescopes, even fairly small amateur instruments, when the planet comes close. Jupiter displays a wealth of changing detail on its face, but what we're seeing is cloud patterns. Only Mars lets us see its land surface, except for when dust storms, frosts, clouds, and fogs temporarily cover regions. Those atmospheric phenomena are exciting to follow, although most are not very prominent to the amateur astronomer save near the edge of Mars or through colored filters that help bring them out. One mighty exception is the larger dust storms. Some appear only as lovely golden patches here and there on Mars. But every so many years there is a planetwide dust storm, and for a while all the surface features of Mars are hidden or made far less conspicuous.

The Martian Polar Ice Caps

Many of the atmospheric phenomena of Mars are intimately related to the most dramatically visible of the planet's seasonal changes: the alternate growth and dwindling of the Martian polar ice caps. The ice caps can be extremely prominent because they are often shockingly white. I especially remember the breathtakingly close conjunction of Mars and the Moon I saw in July 2003 (mentioned in Sight 22), when the ruddiness of the planet and the whiteness of the caps contrasted not only with the yellow lunar surface but also with the blue sky as the Sun slowly rose.

The visibility of either of the Martian ice caps depends on several factors. The first factor that affects how well we see a Martian ice cap is simply our distance from Mars. Much of the time Mars is distant enough for no surface features to be visible in small telescopes, but the planet doesn't have to get much closer and larger for a cap to be glimpsed. Nevertheless, the visibility also depends on a second factor: the angle at which we see the planet. As Earth and Mars move to different parts of their orbits, that angle changes. At some oppositions of Mars, the North Pole is more favorably tilted toward us. At others, it is the South Pole that is tilted our way. At still others, the equator is displayed straight on to us.

The third factor affecting the visibility of a Martian pole is the angle at which the Sun sees the planet—that is, if a pole is tilted toward the Sun, then spring or summer must be occurring for that pole, and the ice cap melts and decreases in size (sometimes the southern cap melts away entirely). If the other two factors are favorable, however (Mars is close and the pole is tilted quite a bit toward us), even a medium-sized telescope may suffice to show a polar cap down to the time it is a tiny—but still flashingly white—speck. Indeed, it is watching the growth or shrinking of an ice cap over the course of weeks and months that is especially thrilling. The planet has its closest oppositions when its southern hemisphere is tilted toward both us and the Sun, and the most memorable transformation of an ice cap we get to see is therefore the shrinking of the southern cap. We even get to see the cap fray around the edge and show the dark lines of rifts in itself as ice melts more quickly from low-lying areas. By contrast, the northern ice cap in winter interestingly develops a North Polar Hood that hides the ice but is itself visible in fairly small telescopes.

A World of Hurtling Moons and Blue Sunsets

Mars is not just the only planet whose surface features we can ever easily see. It is the one whose distance from us varies the greatest and whose visibility of those surface features is provided to us for only relatively short periods of time (a few months) at relatively long intervals (oppositions occur more than two years apart). Even more dramatic is the fact that each opposition improves until the closest one, stunningly better than the others, occurs after either fifteen or seventeen years of waiting for it. (Unfortunately, readers of this book in the year it is published will have some waiting to do: the closest opposition and approach of Mars in almost 60,000 years occurred in 2003, and although the 2005 opposition was very good, another very close one won't occur until 2018. We can keep busy with innumerable other great sights each year—such as the transit of Venus in 2012—while we wait.)

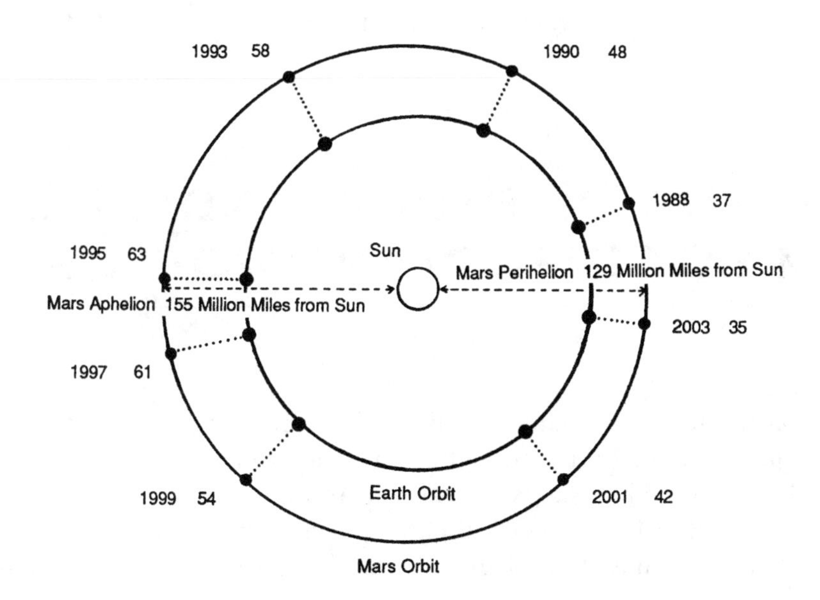

The orbits of Mars and Earth, showing Mar's oppositions. The numbers next to the years are Earth-Mars distances in millions of miles.

Of course, part of the enjoyment of observing Mars is watching its slow but inexorable progress toward our next wonderful opposition with it. I mean here the next time that the two planets pull relatively close together—something that happens at every opposition. But nowadays we can also expect a new spacecraft to rendezvous with Mars every few years. When one arrives, we often get years of eye-opening photos and exciting findings.

For, of course, Mars has always an advantage over other planets when we observe it: we know we are seeing the world most like Earth and a possible home of life. But the similarities of Earth and Mars make the differences all the more thrilling. Planets such as Jupiter or even Venus not only have conditions that are impossibly inhospitable but also so different from Earth's that they are almost incomprehensible (at least as they would be experienced by a human being trying to visit them in person). Mars is enough like Earth that we can relish and marvel at its differences. It is a planet of pink skies and blue sunsets. It is a world with individual volcanoes as big as the state of Arizona and three times taller than Mount Everest. It is a world with a canyon system that on Earth would stretch all the way across the United States and make our Grand Canyon look like a minor ditch. It is a world with two irregular-shaped little moons—one that would look like a very bright star that hardly moved, the other that would look like a little potato flying so fast it would rise in the west, visibly move through the sky, set, and sometimes rise a second time in the same night. This is a world of wilderness wonders that humans with oxygen tanks and thin protective suits will be hiking around in a few decades and perhaps even living on in enclosed cities before our current century is over. This is the world you can see in your mind as your eye carefully scans its intricate surface markings in your telescope. This is the world we call Mars.

Uranus, Neptune, and Other Dim but Important Worlds

One of the guiding principles of this book has been to select sights that are not just of mental interest. Sights, after all, are meant to be seen. And surely shouldn't the best sights be visually arresting?

Quite true. But this chapter is the one place in the book where we really have to bend this rule. All but one of the objects we'll meet in this chapter can be seen either with the naked eye, binoculars, or a very small telescope. But I have to admit that only the first few objects we will discuss here really offer an appearance that is itself unusual and exciting in amateur telescopes. Nevertheless, the other objects do belong in this book. The reason for their inclusion is that they are so important. They are among the important worlds of our solar system, ones we either have already visited with our unmanned spacecraft or will visit in the relatively near future.

When you look at most of these worlds in an amateur telescope, they won't immediately be distinguishable from countless relatively dim telescopic stars. But once you've identified them—with a finder chart and by their motion relative to the fixed stars—their simple sight becomes charged with interest and excitement.

Uranus and Neptune

Actually, the first two worlds in this chapter *are* visually arresting in amateur telescopes, quite independent of any mental interest that they have when you know what they are. I refer to the third and fourth biggest planets in our solar system: Uranus and Neptune.

Uranus shines as "bright" as magnitude 5.7 and can therefore be glimpsed with the unaided eye if you know where to look and have quite dark skies. Neptune is two full magnitudes fainter, so it requires at least binoculars. But it is in telescopes that these giant but awesomely distant worlds begin to show color and size.

I remember how excited I was when I first glimpsed Uranus. I was about fifteen years old and using a 4¼-inch reflector. Way back then, Uranus was in Virgo. The reason I was excited was not just that I had found a planet over 1½ billion miles away all by myself with my little telescope. I was also bowled over by the vivid blue-green color and, when I had increased my magnification a little bit, by the fact that I was able to see the globe of the planet—which was all the more amazing because of how incredibly tiny it looked. Uranus appears no more than about 3.5" across (though, amazingly, this is sometimes enough to make it look bigger than Mars when the latter is farther away than usual).

Neptune is a distinctly harder planet to see well. Although it is similar in true size to Uranus (both of these worlds are about 4 times wider than Earth), Neptune is about a billion miles farther away. Consequently, its apparent diameter is only a little more than 2". On a night when seeing is poor, even a large and well-collimated telescope cannot provide an image of Neptune that will display its minute disk. Furthermore, even a person who is sensitive to color at low light levels needs a fairly large telescope to start seeing a blue-green tint to the dim Neptune. In short, with Neptune the quest for most amateur astronomers is to see any disk or color. When you succeed to any extent, however, the thrill is considerable.

The *Voyager 2* spacecraft is the only probe to have ever passed near Uranus and Neptune. It closely imaged the meager rings and many of the moons of these worlds. Uranus rolls around its orbit on its side and usually has less activity in its clouds than Neptune. Neptune sometimes sports a Great Dark Spot (big but much smaller than Jupiter's Great Red Spot), and its moon Triton is one of the larger and most fascinating in the solar system.

The Challenge of Pluto

After you've seen Uranus and Neptune, there is one more, much farther, much fainter planet left to locate and see, right? Until a few years ago, most authorities would still say that, yes, Pluto is a planet. They would say this despite the fact that Pluto was first observed in the 1970s and was determined to be only about half the size of Mercury and smaller than our Moon. They would say this even though Pluto's small size and highly inclined and elliptical orbit resembles more and more the Kuiper Belt Objects that started being discovered in the 1990s. After all, ever since its own discovery in 1930, Pluto had been considered a planet in the textbooks and by the public. But then in 2005 a key event occurred: a team of astronomers discovered a Kuiper Belt Object larger than Pluto. Should the new world be considered the tenth planet? Or should Pluto be demoted from planethood and simply called a Kuiper Belt Object? In 2006, a committee of the International Astronomical Union (IAU) voted to redefine the term *planet*, leaving Pluto out.

The IAU decision remains controversial. Whatever category Pluto ends up being put into, the fact remains that it can be glimpsed in 8-inch or 10-inch telescopes. The larger but much more distant new Kuiper Belt Object can be seen only in one of the world's best and largest telescopes.

Let me stress, however, that Pluto is a real challenge even for amateur astronomers with a fairly large telescope (though a few of us have seen it, under excellent conditions, in telescopes smaller than 8 inches in aperture). You must have dark skies and a detailed finder chart, for Pluto is too small and far away to appear as more than a starlike object, and its brightness is 14th magnitude. Even with an appropriate finder chart, you should really observe what you think is Pluto on at least two different nights—to make certain that it has changed its position relative to the stars.

The best place to get finder charts for Uranus, Neptune, and Pluto is the astronomical magazines and their respective Web sites (though you can find such charts elsewhere, such as in *Astronomical Calendar*).

The Asteroids

What the partly icy Kuiper Belt Objects are to the outer solar system, the rocky asteroids are to the inner solar system—countless thousands of objects that are mostly found in a belt just beyond the planets of that section of the solar system. A few of the Kuiper Belt Objects can come farther in toward the Sun than the orbits of Neptune and Uranus. Great numbers of the asteroids cross the orbits of the inner planets Mars, Earth, Venus, and even (in just a few cases) Mercury. Astronomers are anxious to discover all the asteroids that have any potential to collide with Earth. We still don't know whether we could divert an asteroid from its path if it were heading toward Earth. (Explosions set off on one side of the asteroid to change its direction of motion could produce the unwanted result of breaking the asteroid to pieces, which would then rain even more widespread destruction on Earth when they arrived here.) One thing is certain, though: the more time we have to give an asteroid a deflecting nudge, the farther off its original course it will be when it enters our vicinity.

Every few years one of these small asteroids comes close enough to Earth to be capable of being seen for a few days in amateur telescopes. If your telescope is large enough and you can locate the object, it certainly can be an exciting sight because you may be able to detect that it is changing position in a matter of minutes (or even, in the most extreme cases, seconds). Do realize, however, that observing these near-Earth asteroids is a challenge beyond the abilities of most beginners (not least of the difficulties is the fact that they are so dim—typically *fainter* than about 11th magnitude).

The asteroids that present the best opportunities for observation, especially for beginners, are the biggest and brightest ones.

The asteroid that can become the brightest is Vesta. It is occasionally bright enough to glimpse with the unaided eye at about magnitude 5.5. Vesta gets its brilliance mostly from being unusually reflective for an asteroid. Much of the time, however, the brightest asteroid is the one that is by far the biggest: Ceres. Ceres measures almost exactly 600 miles across and was the first asteroid discovered. The second was Pallas, which reaches 7th magnitude less frequently than Ceres. When Ceres was discovered on the first day of the nineteenth century, January 1, 1801, astronomers thought it was the long-sought "missing planet" between the orbits of Mars and Jupiter. But in March 1802 Pallas was found. The next asteroid discovery was Juno, a much smaller asteroid that nevertheless occasionally comes close enough to shine as bright as magnitude 7.5. Then, Vesta, the fourth asteroid, was found. And astronomers started talking not about a new planet but about "minor planets" and "asteroids" in an "asteroid belt."

The word *asteroid* literally means "starlike," because even in large telescopes asteroids appear as points of light. Asteroids blend in among the stars in our sky. Nevertheless, they move relative to the stars and remain a throng of great interest. I say "throng" because many thousands are now known and hundreds of thousands exist that are more than a few hundred feet across. Rocky bodies that are smaller than this size are usually called **meteoroids**.

Where can you find the positions of the brightest asteroids when they are well placed? Again, in the astronomical magazines and *Astronomical Calendar*. But in this case there is another very valuable source to consult: the wonderful Web site for satellites called Heavens Above (www.heavens-above.com) also has finder maps and information for locating about a dozen or so of the brightest asteroids at any given time.

Sight 45 A Colorful or Otherwise Striking Double Star

One of the great and wonderful secrets of many stars we see with the unaided eye is that they are not individual stars. A **double star** is a star that upon closer inspection—usually with magnification and greater light-gathering power—proves to be two or more stars.

The exquisite precision of two diamond points of light, one usually brighter, with their rays touching each other, their hearts of light a hair-breadth apart, is breathtaking. But what's even better is that many double stars are not just duos of white diamonds. They are also pairs of sapphire with topaz, emerald with ruby, even combos of tints found only in these stellar jewels, not in the gems of the earth.

Yet, for all their beauty, double stars are less observed than almost any other kind of deep-sky wonder. Why is this?

Observation of double stars used to be more popular among amateur astronomers. Then, about twenty-five years ago, telescope manufacturers started using Dobsonian mounts to make it less expensive to produce

The double star Albireo (see also the color insert).

telescopes with large aperture. The selling point of these "fast" (low focal-ratio) telescopes was not necessarily sharp images but tremendous light-gathering power that enabled amateur astronomers to see vastly greater numbers of extended deep-sky objects such as nebulae and galaxies. That was all well and fine (indeed, very fine!). But there is still a need for telescopes that produce razor-sharp images to help us study the planets—and double stars.

Let me stress from the start, however, that many of the thrills of double stars don't require very high magnification or perfect optics. Although one of the interesting challenges *is* to split the tightest pairs of stars that you can with your aperture. (The greater the aperture, the greater its inherent resolution or resolving ability—that is, ability to see fine details, including the narrowest gaps between two members of a double-star system.) As we'll see in this chapter, it is often not the tightness but the colors and the magnitude differences (or similarities) of double—and triple and quadruple—stars that makes them beautiful.

Defining Double Stars

First, we need to define the several varieties of double stars. A **binary** is a double-star system in which the member stars orbit each other. If one star hides the other or both alternately hide each other as seen from Earth, then the pair is known as an eclipsing binary (the most famous example is Algol—see Sight 29). *Spectroscopic binaries* and *astrometric binaries* are double stars that can't be split visually by any telescope but whose duplicity (doubleness) is discovered by study of the spectrum of the starlight or by a wobble in the motion of a star (it is being tugged by its companion's gravity) as we observe it over the years. A *common motion pair* is made up of two somewhat more widely separated stars that, although not orbiting around each other, are traveling through space together with the same speed in the same direction. An **optical double** is a pair of stars that may appear rather close together in the sky but are at vastly different distances from us—in other words, one star just happens to lie along almost the same line of sight from Earth but has no special physical connection with the other star in space.

Although *double star* is the blanket term for all star systems that are not single, we sometimes refer to a star system with more than two members as a **multiple star**.

Other than the magnitudes of the component stars, there are several statistics we want to know about a double star. First is the *separation*. It is usually given in seconds of arc. Second is the *position angle*, which is the direction of the *companion* (almost always the fainter star) from the *primary* (almost always the brighter star). The position angle, or PA, is normally given in degrees like azimuth (0° is due north, 90° due east, 180° due south, 270° due west, and so on through 360° until we are back to due north).

Naked-Eye and Other Wide Double Stars

A few double stars are separated widely enough to split with careful naked-eye scrutiny, depending on how sharp your vision is. The 2nd-magnitude Mizar, the star at the bend in the handle of the Big Dipper, has its 4th-magnitude companion, Alcor, about 11' away from it, and most people can see both with the naked eye as long as sky conditions are reasonably good. (If you use mod-

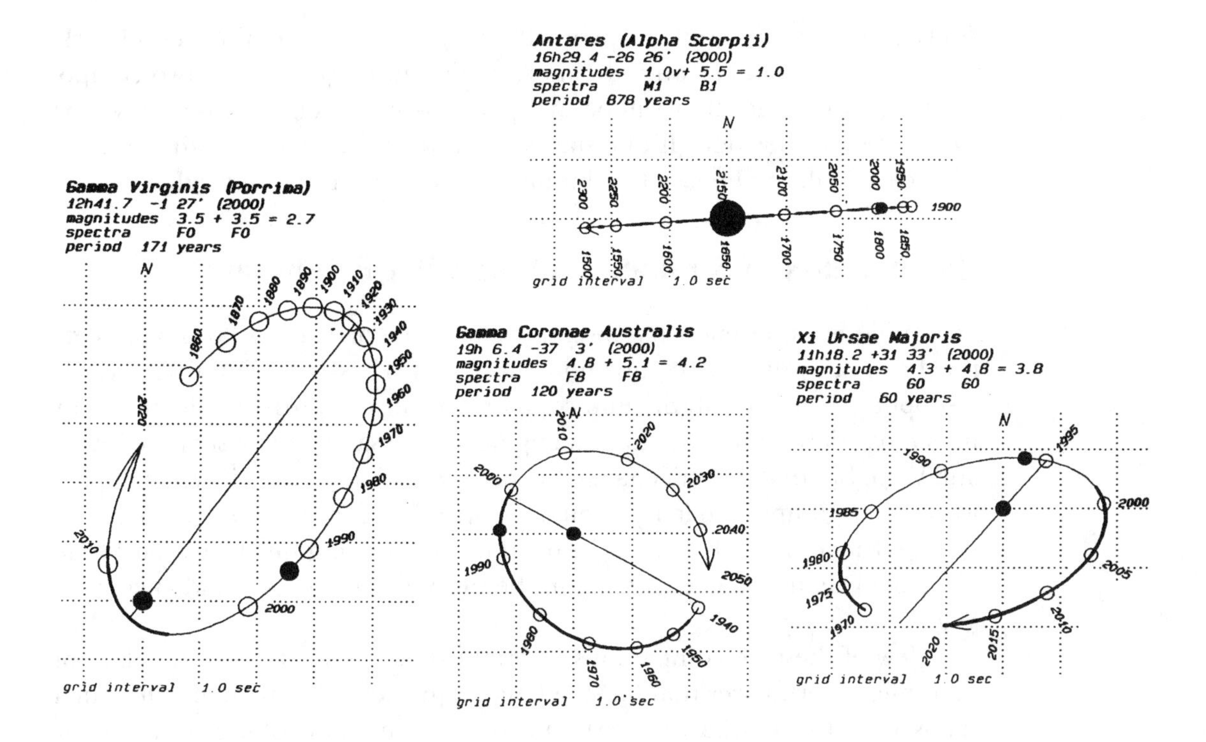

The orbits of some important double stars.

erate telescopic magnification on Mizar, you'll find that it splits again into a fine double.) More difficult for the naked eye to split are stars such as Alpha Capricorni and Alpha Librae. A real challenge for most people (separation 3.5' apart) is Epsilon Lyrae, near Vega—you might see it as elongated, but your vision has to be rather acute to split it cleanly. (The split is easy in binoculars, however, and if you use about 100× to 150× on the two you discover—well, we'll get to what you discover in just a moment.)

The ultimate double-star split for the naked eye seems to be the faintest of four stars in the head of Draco the Dragon. This is Nu Draconis, also known as Kuma. In this system, two magnitude 4.9 stars are 63" apart. When I challenged readers of *Sky & Telescope* to see if they could "conquer Kuma," only one person reported being able to do it—when he was younger, and in a marginal way.

Of course, any of these stars can be split very easily with the use of a little bit of magnification. Use the lowest magnification you can on a telescope to see the two stars of Nu Draconis blaze bright as white twins still rather close together.

The case of Nu Draconis brings us to two important points to make about observing double stars. First, you will generally want to use the lowest magnification with which you can just barely separate the two components of a double-star system. By doing so, you are packing more majesty into a tighter package and can be awed by the breathtaking narrowness of the waist of darkness between them. Second, the more similar in brightness the two components are, the easier they will be to separate—or, rather, the easier it will be to see the dimmer one. If one star is much brighter than the other, then it will be difficult to perceive the dimmer in the glare from the brighter.

Components of Similar and Dissimilar Brightness

For a unique experience, try observing a double star, such as Polaris or Vega, that has a vastly dimmer companion star near it. These companions are really just optical doubles with the dim component—in our present millennium or millennia—just happening to lie along nearly the same line of sight as Polaris and Vega. But that doesn't change the beauty of what you see. These companions can be glimpsed in fairly small telescopes. But I love the view of Vega and its little friend in a 10-inch telescope: the companion glints like a hard little piece of gleaming steel trembling in the vast wash of sapphire radiance from Vega.

A few of the true companions of bright stars are so close to the glare of their mighty primaries that the challenge is just to see them at all. One such star is Sirius B—the (Earth-sized) white dwarf companion of the brightest star in the heavens. In the opening decades of the twenty-first century, the apparent separation is growing between the famous Dog Star and its companion (playfully nicknamed "the Pup"). Even so, spotting Sirius B will be a true challenge for observers with medium-sized telescopes. In summer, orange Antares has a surprisingly bright companion—a magnitude 5.4 star. But it is located only 3" from its 1st-magnitude primary. A good telescope as small as 4 inches can show Antares B, but you'll need a night of really good seeing, especially if you live in midnorthern latitudes where Antares is always low in the sky. The effort is worth making, though, because the companion of Antares can appear vividly green.

Seeing double-star components of greatly different brightness is fascinating. But there is also much to say for pairs of similar or identical brightness—twin wonders. Gamma Virginis (also known as Porrima) is such a star. Its combined brightness is 2.2, but its components are both magnitude 3.5 stars of F0 spectral class (because they are a true binary, with both stars only 27 light-years from us, they really must be suns of identical luminosity and color). Unfortunately, the pair is too close together for most amateur telescopes to split for the next several years. By 2011, however, they will have separated to 1" apart and by 2015 to 2' apart. What do they look like when you can split them? In the words of Robert Burnham Jr., they are like "twin headlamps of some celestial auto, approaching from deep space."

Multiple Stars

A few paragraphs earlier, I stopped short of telling what happens when you apply a lot more magnification to the wide double star that binoculars (or sharp naked eyes) show Epsilon Lyrae to be. Now I'll reveal the secret: each of these two white stars of nearly identical brightness (magnitude 4.6 and 4.7) further splits into a similar pair—one white pair of magnitude 5.0 and 6.1 stars 2.6" apart, the other white pair of magnitude 5.2 and 5.5 stars 2.3" apart. Thus Epsilon Lyrae is one of summer's most delightful telescopic sights and is known to amateur astronomers everywhere by its nickname: "the Double Double."

The only quadruple star that is perhaps even more fascinating than the Double Double is one that is so partly because of its location: the quadruple (actually at least sextuple but the fifth and sixth stars are very dim) star Theta Orionis shines in the very heart of the mighty Orion Nebula. See Sight 46 for discussion of Theta Orionis and Sight 28 for discussion of other multiple stars in the region of Orion's Belt and Sword.

The triple star Beta Monocerotis is marvelous by virtue of the tightness of its trio, the similarity of the stars' colors, and the relative similarity of their brightnesses. Most people see these stars as white. The B and C stars shine at magnitude 5.2 and 6.1, respectively, and are only 3" apart—with the A star shining at magnitude 4.7 and just 7" from the B star and 10" from the C star.

The Many Colors of Double Stars

There are few celestial objects or phenomena that show prominent color. And even when you look at bright stars in telescopes, you may find that their hues are subtle. That is, unless they are parts of a double star.

The eye-brain system's perception of an object's color depends crucially on the color of objects near it. I've already mentioned in this book how close conjunctions of bright planets with each other or with stars can often greatly intensify the colors of both objects that are perceived. This effect is nowhere more wonderfully exemplified than in the colors of many double stars.

The most famous of colorful double stars is unquestionably Beta Cygni—better known as Albireo. This star marks the beak of Cygnus (the much brighter Deneb is the tail of the Swan). To the naked eye it is a modest magnitude 2.9 point of light. But very low magnification—even that of many finderscopes and binoculars—begins to split the point into a magnitude 3.1 and 5.1 pair. A small- to medium-sized telescope at low power then gives a comfortable split and unfolds for us, as if by miracle, what the veteran deep-sky observer James Mullaney has called "one of the grandest sights in the entire heavens." We see in the midst of a Milky Way field plentiful with stars the rich golden primary and pastel but intense blue companion that makes up Albireo.

Albireo is often called the most beautifully colored double star for small telescopes. But there is a tremendous amount of subjectivity in such assessments. As much as I love Albireo, I'm amazed that only in recent years have astronomy writers begun to recognize a triple star in Cygnus that I think is a wonderful alternative beauty. I refer to Omicron-1 Cygni. It is one of two stars (the other is Omicron-2 Cygni) about a degree apart that are only a few degrees northwest of Deneb. Turn your medium-sized telescope at low magnification on Omicron-1 Cygni and you see a magnitude 3.8 gold star a full 338" away from a delicately green star of magnitude 4.8. And then you notice that 107" from the gold primary is a magnitude 7.7 star with a beautiful hue. Some people call this hue blue—but to my mind and eye it is purple. I first stumbled upon this beautiful trio—by myself—one night back in the late 1970s. The next day, I was walking in bright June sunshine when I noticed a variety of small wildflowers growing a few feet from me in a roadside field. Their color literally leaped out at me and I almost gasped—for it stunningly brought to mind the color of the magnitude 7.7 component of Omicron-1 Cygni.

Why do different people see and name the color of a particular star differently? Part of the answer is physiology: for example (and it is just one example), some people are more sensitive to the red end of the spectrum, others to the violet. Another part of the answer is color vocabulary, which varies greatly from person to person. As evidence of this, but also as proof of the variety of double-star colors and the passion they inspire, here is a sampling of the names given to the tints of various pairs: pale garnet, cream-colored, port-wine tint, olive, tawny, greenish-yellow, bluish-gray, ashy, topaz, crocus yellow, rosy, reddish-violet, warm gray, clean yellow, flushed white, intense blue, emerald-green, greenish-gray, pale orange, clear blue, grayish lilac, smalt (deep) blue, pea green, cerulean blue, dull red, pale orange, sea green, light apple green, cherry red, indigo blue, sapphire blue—and so on and on!

Tips on Observing Star Colors

Before I conclude here, I'd like to make a few points about how best to see strong star colors.

First, star colors depend very much on the size of your telescope's aperture. In the cases of some double-star colors, having a large telescope is actually undesirable. For instance, a magnitude 3.1 and 5.1 pair like Albireo probably show their most intense color in about a 4-inch or 6-inch telescope. Use a much larger instrument and the light-gathering power will make the two stars brighter—too bright, for the hues will appear less saturated. Of course, the dimmer a double star, the larger the aperture that will be needed to give strong color.

It's also true that with larger telescopes the lowest magnification that can be achieved may be too high—high enough to separate a particular double star farther apart than is ideal. The key is to keep the pair separated but extremely close together to concentrate the splendor—and intensify the difference in the two colors that the mind perceives.

A trick some people use for intensifying the color of a bright double, such as Albireo, is setting the images slightly out of focus. Then the color is spread out across disks of light. It does seem to make the colors easier to study. But I think of this as a pleasant additional experience to the in-focus view, which offers the stars as intense, trembling jewel-like points.

The Great Orion Nebula

Think for a second of all the different kinds of astronomical objects beyond our solar system that are interestingly visible in telescopes: bright stars, colorful stars, variable stars (including novae and supernovae), double stars, open star clusters, globular star clusters, emission nebulae, reflection nebulae, planetary nebulae, dark nebulae, supernova remnants, and spiral, elliptical, and irregular galaxies. This list could go on and on if you wanted to delve into subcategories of the different types. And yet, the vast majority of amateur astronomers would agree as to which individual object beyond our solar system is the grandest and most intricately beautiful in their telescopes. That object is M42, the Great Orion Nebula.

A Translucent Green Fan of Glory

We've already studied the Belt and Sword of Orion back in Sight 28. Together, they constitute as rich and splendid a small region of stars as any in the heavens. But we saved for this current chapter the crowning glory of the

The Great Orion Nebula.

region, and it is not a star or a star cluster. It is a **nebula**, a cloud of interstellar gas and dust. We study other wonderful nebulae in this book—including the great diffuse emission nebulae of the summer, objects such as the Lagoon Nebula and Trifid Nebula, back in Sight 32. But M42 is the brightest of all nebulae visible from Earth.

Under good sky conditions, most people will see the nebula as only a strange blur of light around Theta Orionis, the 4th-magnitude middle star in the Sword. But turn even a weak pair of binoculars on that star and the wonders begin to unfold.

In binoculars, M42 becomes a puff of mysterious glow around the star (a star that turns out to be double in binoculars). You begin to see that there is much more to what seemed only a gleam in the Sword to the naked eye. With even a small telescope, the nebula expands and brightens further. By the time 80× or 90× is used on an average 4-inch or 6-inch telescope, the main part of the nebula fills most or all the field of view—and you are confronted with one of the most awesome sights in all of astronomy.

In a 6-inch or 8-inch telescope, the Orion Nebula is so beautiful and intricate it staggers the mind. Even in the city or on a night of the Full Moon, its brightest region is easily visible and so intense that its details are hard-edged. But wait until you get a look from a country site. Then you see it blossom out from the center into a vast giant translucent fan of green radiance. Within this fan, wisps, streamers, even curls of light stretch far out into space with numerous stars, bright and faint, twinkling through the awesome diaphanous luminous veil. Some observers compare it to a giant bloom, others to a glowing hand, still others to an impossibly large ghostly moth or bat. The hard bright northern edge of the nebula extends in the most distinct streamers—which themselves sometimes seem like widespread wings.

No other deep-sky object is so awesome that observers discuss special ways to try to grasp, to try to take in, all its details, all its glory. Some observers suggest placing M42 just outside your field of view and letting Earth's rotation slowly bring it, part by part, into your vision. As one bright and intricate part after another creeps into view, you find yourself gasping and marveling again and again. In the several minutes it takes for the whole vast array to drift slowly by, you drink up detail after detail of beauty. Then with a deep breath, you move the telescope back to begin the process all over again—and again. Whether you observe it this way or keep it (by hand or clock drive) centered in your eyepiece, you will find yourself transfixed. This is the one deep-sky object that even the most initially casual and impatient observer does not leave. I am quite serious when I say that you may easily find yourself spending an entire hour at a time studying this nebula, swimming in its beauty, reveling in its glory.

The Structures of Glory

There is truly no end to the wonders you can observe in M42. It is in this respect like a painted masterpiece—which is indeed a very apt comparison. But to enjoy more deeply a great painting one does need to study the major parts and understand how they are related to one another—the painting's "composition"—as well as admiring some of the boldest individual brushstrokes.

Our scrutiny of M42 begins with its brightest area, named for the sixteenth-century astronomer Christiaan Huygens, who performed some of the first detailed examinations of the nebula in the excellent telescopes he built. This Huygenian Region is something that even M42's sole rival—southern hemisphere observers' great Eta Carinae Nebula—cannot match. No part of any diffuse nebula glows so bright. When I first got a 13-inch telescope about twenty-five years ago, one of the things that amazed me was the brilliance of the light pouring out of the eyepiece when I had the telescope centered on M42. I held my hand in front of the eyepiece and was stunned by the fact that my hand was plainly green in that radiance. The veteran observer James Mullaney tells the story of focusing a big 30-inch observatory telescope on M42 and actually being able to project a full-color image of the nebula from it onto a sheet of glossy white paper.

The second thing you may notice after recovering from being initially stunned by the brightness of the central region itself is an incredible tight little pattern of four stars within it. Not far within the apex of the fan of the nebula sits this possibly most striking of all quadruple (actually sextuple) stars. It is nicknamed the "Trapezium." We'll study it more closely in a second. But next your gaze is pulled from the Trapezium and the nebula's apex to yet another nearby marvel: the deepest, darkest, and hardest of the dark bays that edge the nebula. This dark bay that eats into the central region of the Great Nebula is called the Fish's Mouth (a name given to it by one of the two great deep-sky guidebook writers of the nineteenth century, Admiral William Henry Smyth). The Fish's Mouth helps separate the Huygenian Region of M42 from a comma-shaped detached part of the nebula that gained a separate Messier designation: M43.

Before you begin your observing session's potentially endless tracing of the nebula's intricate branches, bays, rifts, streamers, and filaments, you should check out an amazing aspect of the structure that is visible throughout much of the bright part of the nebula. We are assuming you have at least a 6-inch or an 8-inch telescope and are now using somewhat more than your lowest magnification (about 100× is advisable). Do you see the mottling in the nebula's glow? This mottling was first beautifully discussed by the great nineteenth-century astronomer Sir John Herschel. "I do not know how to describe it better," wrote Herschel, "than by comparing it to a curdling liquid, or a surface strewed over with flocks of wool, or to the breaking up of a mackerel sky."

From a fascination with how bright and detailed M42 is, you will turn to the question of how large an area it covers—how far out can you detect the faintest hint of its glow? For one thing, can you trace beyond M42's far-extended wings a huge loop of dim glow that makes a full circle with the nebula and connects it with the Iota Orionis (the bottom and brightest star of the Sword) complex of nebula and cluster? (See Sight 28 for more on Iota and its surroundings.)

The Trapezium and the Colors

Now let us turn to two special topics that are of utmost interest to the observer of M42: the multiple star within it and the colors—not just one color—that glow from it.

The Trapezium—the stars that make up Theta-1 Orionis—is an uneven quadrilateral of stars that can be separated from one another by even rather small telescopes. The stars are designated A, B, C, and D, and the sides of the geometric figure they form measures 9", 19", 13", and 13" long. Large amateur telescopes reveal two more components, E and F. The A, B, C, and D stars are lettered not in order of brightness but in order of right ascension (west to east) and are of magnitudes 6.7, 7.9, 5.1, and 6.7, respectively. B and A vary in brightness, however. B is known to variable-star observers as the eclipsing binary BM Orionis. It varies from about 7.9 to 8.6 every 6.5 days. The variability of B has long been known but not so A's. A's marked and regular variability wasn't noticed until 1973, when German astronomer Eckmar Lohsen did so. Trapezium star A dims by a whole magnitude, in an eclipse that lasts for 20 hours and has the star at minimum light for 2.5 hours. But these eclipses are not extremely frequent. They recur at intervals of sixty-five days—almost exactly 10 times longer than B's period. Only about a dozen of A's eclipses had been properly observed before 2005.

Much larger telescopes are needed to detect the 16th-magnitude stars near the Trapezium dubbed G and H. On the other hand, Theta-2 Orionis is easy to see but often overlooked by telescopists. Theta-2 is 2' 15" away from Theta-1 and consists of magnitude 5.1 and 6.4 stars that are 52" apart. The brightness of magnitude 4.6 and 4.8 Theta-1 and Theta-2 Orionis combined is 4.0—bright enough to make observing M42 with the naked eye a bit tricky for most people, even though the nebula's total brightness is usually listed as 2.9.

From the challenge of trying to glimpse the faint extra components of Theta-1 Orionis, we turn to the quite different challenge of seeing extra colors in M42. Unfortunately, some people's color vision at night is such that they have trouble seeing even the green of the nebula, even with large telescopes. This is amazing to a lot of us, who can detect that hue so easily as the size of the telescopic aperture increases from 4 inches to 6 inches to 8 inches. The characteristic lovely green is produced by emissions of doubly ionized oxygen in the nebula (more on this in a moment). But many of us begin to see one or more other colors as the telescope size increases even more. James Mullaney says that apertures as small as 5 inches will show "faintly glowing pink areas." Pink makes sense because this would be the glow from hydrogen gas, by far the predominant ingredient of the nebula. My own experience was to first really notice what I would call pale purple or purple-red parts of the nebula with a 13-inch telescope (then somewhat smaller telescopes once I knew what to look for). This color is most prominent out along the streamer that forms the sharp northern edge of the eastern half of the nebula. But what causes the purple tint? I've speculated in my book *The Starry Room* (see the sources at end of this book) that the reason some of us see this purple is that it is actually violet caused by an emission at the extreme violet end of human vision.

The reasoning went as follows. My good friend, the veteran observer Steve Albers, had glimpsed a pale violet in parts of some Northern Lights displays and then seen it show up plainly in photographs (most commercial film is more sensitive than the human eye to the far violet end of the spectrum). This must have been from auroral emissions of ionized molecular nitrogen at a wavelength of 390 nanometers that occur in the extremely rarefied air high in Earth's atmosphere. Books sometimes say that the shortest wavelength visible to the human eye is 400 nanometers, but this value is certainly exceeded by some individuals. There is also certainly ionized nitrogen in the Orion Nebula, and the gas of the nebula is certainly rarefied—in fact, almost unimaginably tenuous. Are Albers and I detecting this far-violet glow in the Orion Nebula? Or are we just seeing a purple caused by a blending of red hydrogen emissions with blue light reflected off the gas from nearby stars? I'm even more puzzled when I think how I (and I believe many people) often see the much-fainter companion in a double-star system shine with a sort of wine-red or purple-red light.

Regardless of what hues I perceive and what their explanation is, what colors do *you* see in the Orion Nebula?

M42 and Other Clouds of Starbirth

Of course, while it is possible to enjoy M42 simply for the rich feast of visual impressions it provides, we can't help being curious about the physical nature of this most spectacular looking of all the deep-sky objects.

The truth is that M42 is the closest and brightest hotbed of star birth in all the heavens. There are two major kinds of glowing **diffuse nebula**. A *reflection nebula* glows merely by reflecting the blue-white light of hot, young stars near it or in it. An *emission nebula* glows on its own after being excited to fluorescence by the ultraviolet radiation of even hotter stars near it or in it. The elusive nebulosity associated with the Pleiades is a reflection nebula, because the Pleiades aren't quite hot enough to ionize it. But the glow of M42 and of summertime favorite nebulae such as M8 (the Lagoon Nebula) and M20 (the

Trifid Nebula) is caused largely by their gases actually emitting radiance of their own after being charged up by the ultraviolet radiation from the extremely hot and young stars around and inside them.

"Young" is a key word here. Some astronomers have estimated that the stars of the Trapezium burned their way into visibility from out of M42's gases not billions or millions but as recently as 20,000 years ago. Whether or not this is true, studies of the nebula at other wavelengths, especially from space telescopes, reveal a still-hidden cluster within the nebula and numerous other wonders of stellar infancy. "Proplyds" are "protoplanetary disks" (solar systems in the making), and astronomers have found proplyds aplenty in M42. There are many other mysteries of the Orion Nebula for astronomers to explore. Especially haunting is evidence of planets ranging freely, bereft of their solar systems, through the nebula.

The Orion Nebula does indeed look like a swirling mass of turmoil in which stars could be born. But what's awesome is that this vast and furious cloud creation is frozen to seeming stillness by the still greater immensity of time and the roughly 1,600 light-years that separate us from it. The impression you're likely to take from an evening of observing the Orion Nebula is of mighty power, action, and glory all calmed to (and I choose my strange, paradoxical words deliberately here) breathtaking peace.

A Rich Open Cluster

Sight 47

It's fundamental. If there is one purpose that an astronomical telescope is for, it is to enable one to see a richness of stars sprinkled upon a background of night sky. We've already explored the topic of where the overall starriest fields of view are for observers using their naked eyes, binoculars, and wide-field telescopes (see Sight 31). But what if we are willing to increase our magnification and home in on even more intense richness of stars? Think of a bursting purse full of diamonds of various sizes spilled into a heap on a cloth of black velvet and lit so as to appear to be shining by their own light in an otherwise jet-black room. But in this case the diamonds are stellar, the rich dark cloth is night, and those jewels really do shine on their own: you are staring into the crowded midst of a rich open star cluster.

The Lure of Rich Open Clusters

Back in Sights 25 through 27, we explored the star clusters that lie within about 600 light-years of Earth. These, we saw, are the only ones close enough to not only appear big to the naked eye but also to have at least a few of their brightest stars visible to the naked eye (M44, the Beehive Star Cluster, was borderline in the second of these respects).

But if we move out farther from Earth and equip ourselves with a telescope, this situation changes dramatically. Even just binoculars will show many dozens of star clusters that mostly appear as fuzzy patches of light, maybe with the glitter of a few bright cluster members in them. In contrast, a good medium-sized astronomical telescope can give pretty interesting views of literally hundreds of star clusters. The great majority of these are not globular star clusters, which are mostly a few tens of thousands of light-years away (but fascinating—and featured in the next chapter). Instead, they are **open clusters** (also known as *galactic clusters*).

Open clusters earn their name from the fact that we typically see quite a bit of space between the individual stars. This is opposed to our view of globular clusters, which are so much more densely packed that they largely appear as balls of continuous glow with only especially bright or especially outlying stars glittering individually.

Is there a happy compromise between the extremes of globular clusters, which have their stars too bunched together, and open clusters, which have their stars too widely (and thus sparsely) spread? Yes. It is the few brightest and richest of the open clusters. They give us not a hundred thousand stars overlapped into a single glow or a few dozen stars scattered too broadly. They give us a wealth of up to several hundred individually visible sparklers, many bright, hoarded into a single telescopic field of view.

The Double Cluster

Not only do open clusters present us with individual stars. They are themselves wonderfully individual—in the sense of each being different in appearance from every other. They offer a wide variety of numbers of stars, densities, arrangements of stars, and balances of brighter to dimmer stars. Add to this the variety of different appearances they present through different telescopes and magnifications and you realize that open clusters are endlessly fascinating to observe and that there is an original experience to be had in observing any of them.

A corollary of this truth is that the title of most beautiful or spectacular open cluster in the telescope ought to depend on the telescope and, most

The Double Cluster in Perseus.

important of all, on the aesthetic preferences of the observer. And yet, I'm inclined to think that there are a few of these clusters that would be at or near the top of every discerning observer's list of favorites. So I wish to concentrate much of my discussion here on the objects that top my own list of the most spectacular rich open clusters visible in small- and medium-sized telescopes.

I'll begin with the Double Cluster in Perseus.

Imagine your amazement when you look in your telescope's eyepiece and see not just one of the richest and most splendid of all open clusters but a pair of them, side by side. But let's back up for a moment.

On any clear, moon-free autumn evening, the Double Cluster in Perseus is easily visible to the naked eye (though not close enough to Earth or open enough for the naked eye for us to see any of its individual stars). In even just a moderately dark sky, the paired clusters look like an elongated little fuzzy barbell of light to average unaided vision. This sight was known in ancient times, but it wasn't until the seventeenth century and the very threshold of the age of the telescope that the two parts of the barbell received Greek-letter designations as if they were stars: h Persei and Chi Persei. In the nineteenth century, they were classified properly as deep-sky objects and given the New General Catalog designations of NGC 869 (h Persei) and NGC 884 (Chi Persei).

What will matter to you on a dark autumn night is not so much their classification but their appearance in your telescope. They are in a convenient location—about midway between the main patterns of Perseus and Cassiopeia. If you leave off the slightly dim star at the eastern end of the famous zigzag figure of Cassiopeia, that end of the pattern points to the Double Cluster. And then you put a low-power eyepiece in your telescope and look in.

If you're like most people, you will gasp or murmur a little. The low-power eyepiece is necessary to give a field a little over a degree wide to fit in the two roughly half-degree-wide clusters. That field now contains twin—well, not twin but kindred—piles of overflowing wealth of jewel-like stars. A few dozen stars stand out bright enough in each to be seen in even small telescopes. With a fairly large amateur telescope, a total of a few hundred stars, true cluster members and nonmember foreground stars, is detectable in the single field of view that holds both clusters. Note both the similarities and the differences between the two clusters—do you see the reddish star that stands out near the center of one of them?

Just the sight of the Double Cluster is spectacular enough, but when we hear how distant these paired objects are, we experience a new level of awe. These clusters lie 7,000 light-years from Earth. They are not even located in our spiral arm of the galaxy but in the next arm out from ours, the Perseus Arm. There are at least ten stars in the Double Cluster whose true brightness rivals that of our most famous blue giant and red giant, Rigel and Betelgeuse.

M11, the Wild Duck Cluster

Everyone will have their own favorite cluster in the telescope (or several favorites to suit different moods). My favorite telescopic open cluster is perhaps the only one visible from the northern hemisphere that can really compete with the Double Cluster in raw splendor: M11.

M11 is only a single, not a double, collection of stars. Its total brightness is considerably less than that of the Double Cluster. But it is much richer than either member of the Double Cluster. On a 1-to-5 scale of cluster richness (concentration) based on that of the great astronomer Harlow Shapley, the Pleiades and Hyades rate a 1—the least concentrated on the scale. The Double Cluster rates a 4 (h Persei) and 3 (Chi Persei). But M11 rates a 5—the greatest degree of concentration possible for an open cluster.

Then there is the number of stars in these clusters. Counting very faint members of the clusters, M11 has about twice as many total stars as the two clusters of the Double Cluster combined. And M11's stars are concentrated into less than half the diameter of either of those clusters!

Let me describe the experience of "sneaking up" on the splendor of M11 by starting with lowest magnification and light-gathering power and slowly increasing both.

M11 is bright enough to see with the naked eye in a really dark sky though somewhat hard to glimpse that way because it lies in front of one of the very brightest star clouds of the Milky Way, the Scutum Star Cloud. With binoculars or a finderscope, however, you can easily locate M11 as an intense spot of light in the midst of the star cloud. Do you have a small telescope? Look through it at low magnification and the spot turns into a strange luminous triangular patch with its brightest stars glittering in it. But the real thrills begin when you turn about a 6-inch telescope on M11.

There before you are the brightest stars of the cluster, arranged in a flying wedge—no wonder the nineteenth-century deep-sky observer and writer Admiral William Henry Smyth coined the nickname "Wild Duck Cluster" for M11. But what's most dramatic is that the leader of this flight of stellar ducks or geese is the decidedly brightest point of light, an 8th-magnitude yellow star. And that all around the V-shaped flock are not just dozens and dozens of fainter individual stars but also a beautiful luminous mist. The first time I saw this view in a medium-sized telescope I described it to myself as an avalanche of stars throwing up a cloud of nebulosity. The avalanche is, of course, fanning out from the brightest star, which itself is at the apex of the V or triangle. But the beautiful mist behind the individually visible stars is not nebulosity. In a dark sky, a larger telescope—an 8-inch or a 10-inch—at slightly higher magnification reveals that the mist is entirely composed of faint stars. Hundreds of faint stars. To say the view is breathtaking is an understatement.

Other Superb Open Clusters

There are enthralled and detailed descriptions that could be passionately provided for many, many other open clusters that are beautiful in their own special ways in the telescope. But I can offer only the most lightning-quick surveys here of a few of the most outstanding open clusters for amateur telescopes. (Remember that this listing is in addition to the Double Cluster and M11 and doesn't include those loose giants the Hyades, the Pleiades, the Alpha Persei Association, the Beehive, and the Coma clusters.)

Autumn

M52 and NGC 457 are just two of the many lovely open clusters of Cassiopeia. NGC 457 is known as the Owl Cluster or E.T. Cluster (as in *E.T.*, Steven Spielberg's movie about an extraterrestrial) because it looks like a figure with long downward hanging wings or arms, and it sports two stars for eyes. (One of the eyes, the brightest, is magnitude 4.8 Phi Cassiopeiae—an orange supergiant at least 5,000 light-years away whose true luminosity may be greater than that of Rigel.) Two big low-magnification clusters of autumn are M34 in Perseus and NGC 752 in Andromeda (the latter lies right next to the gorgeous wide double-star 56 Andromedae, composed of two orange 6th-magnitude stars). In the dim constellation Camelopardalis, NGC 1502, the Golden Harp Cluster, is bright enough to glimpse with the naked eye—and at low magnification a telescope finds it near one end of the remarkable 2.5°-long string of stars known as Kemble's Cascade (this asterism was discovered in 1980 by Canadian amateur astronomer Lucian J. Kemble).

Winter

M38, M36, and M37 are three bright and rather rich open clusters almost in a row—a row that extends from out of the pentagon pattern of Auriga. M38 and M36 are only about 2.3° apart, and the former forms a sort of cross or letter pi pattern of stars (M38's brightest star is an 8th-magnitude yellow giant). M37 is a little farther southeast and is the biggest, brightest, and richest of the three. Bigger and brighter than M37 (in fact fairly easy to see with the naked eye in dark skies), but less rich, is M35, located in the northern feet of Gemini. (M35 needs to be observed at very low power to look best, but if you up the magnification you may detect the hazy spot of a 6-times-more-distant cluster, NGC 2158, on its fringe.) Winter's great abundance of open clusters continues with the heart-shaped M50 in Monoceros and M41, a lovely low-magnification cluster bright enough to sometimes be glimpsed with the naked eye only about 4° south of Sirius (the brightest star in M41 is a reddish one near its center). Last but not least, just 1.5° apart in Puppis are the wonderfully dissimilar clusters M47 and M46 (M47 is lozenge shaped, brighter, but sparser; M46 is a big, roundish cluster composed of mostly magnitude 10 to 13 stars).

Spring

Open clusters are found mostly near the equatorial plane of our galaxy and thus near the Milky Way band in our sky—but that band is low in the sky in spring. Besides M44 (the Beehive) and the Coma cluster, however, spring offers M48 in Hydra and the rich but very old M67, which shines less than 2° west of Alpha Cancri (a star also known as Acubens).

Summer

M6 and M7 are the two big, bright clusters that glow just 3.5° from each other, near the sting of Scorpius. We discussed them and the Sagittarius open clusters M23 and M25 in Sight 32 (our telescopic tour of the Sagittarius region). But a far less well-known cluster of splendor lies 0.5° south of the wide double-star Zeta Scorpii. This spectacularly overlooked object is part of an amazing north-south complex of stars and clusters curiously named "the Table of Scorpius." Brocchi's Cluster (Collinder 399), located in Vulpecula about a third of the way from Altair to Vega, is better known as "the Coathanger"—because this asterism (not true cluster) at low magnification looks mind-bogglingly just like a perfect coat hanger made of stars. Finally, way up in northeastern Cygnus about 9° north-northeast of Deneb is M39, which looks something like a much farther, much dimmer Pleiades. But M39 is bright enough to apparently have been recorded by Aristotle in ancient times.

A Bright Globular Cluster

Sight 48

In the previous chapter, I mentioned some of the differences between open clusters and globular clusters.

Before we explore together the sight of bright globular clusters in this chapter, however, it would be a good idea to recall a few facts that demonstrate the nobility and importance of these objects in comparison to open clusters.

While there are thousands of open clusters in our galaxy, there may not be greatly more than a hundred globular clusters. While open clusters are mostly confined to fairly near the equatorial plane of the Milky Way, globular clusters hover around the center of the galaxy in all directions and at what are in some cases tremendous distances from the center. While a few most abundant open clusters have up to about a thousand stars, globular clusters have tens of thousands, hundreds of thousands, perhaps up to a million or more stars each. On average, the open clusters we see in the sky are many times closer than globular clusters—in fact, the most distant prominent open clusters (M11 at more than 5,000 light-years and the Double Cluster at more than 7,000 light-years) are no farther than the very closest globular clusters (which might be just under 7,000 light-years away). Considering the great remoteness of globular clusters then, it is not surprising that open clusters can *appear* brighter and larger and can look splashier and flashier (young open clusters do host some of the most massive and luminous of stars—in this respect their flashiness has a legitimate basis). But a globular cluster offers us the special thrill of knowing we are seeing one of the galaxy's greatest citadels of stars and viewing a ball of up to one million suns glowing with innumerable pinpricks of individual stars glittering through it.

The globular cluster Omega Centauri.

A Variety of Globular Clusters

Some lazy observers would say that what I just wrote is indeed what there is to see of globular clusters—a hazy ball of light with maybe some sparkles here and there. But, they would add, that's *all* there is to see. If you've seen one globular cluster, they would say, you've seen them all. Any avid, passionate amateur astronomer who has really bothered to look knows that they are wrong.

Globular clusters certainly tend to appear far more similar to one another than open clusters are to other open clusters. But most of the bright globulars have some unusual, maybe even unique, aspects that set them apart from their siblings. Some of the differences are really quite striking.

For proof of this, you really have to look no farther than the two bright globular clusters in the constellation Hercules. M13 is the globular every telescopic astronomer beyond novicehood knows, a big bright monster. Just 9° away, yet harder to locate in relation to prominent stars, is the overlooked and underobserved M92. At first glance, you might be disappointed after viewing M13 because this second cluster looks far smaller (that's a difference, but not one in M92's favor). But the instincts of your eye-brain system will be sending you urgent messages that this sight is not at all disappointing. What is so beautiful about M92? The most outstanding feature of it is the great extent to which it is concentrated in its middle regions—much more so than M13. It is this blazing core of the relatively small M92 that makes it end up having not that much less total brightness than M13.

There are other engaging differences between M13 and M92. There are major differences in the size and arrangement of what seem like streamers or arms (composed of stars) extending out from different globular clusters. Do you see any color in either of these clusters? Some globulars do display color to many of us in medium-sized and larger amateur telescopes, and there is considerable variety in the golds and blues and how they are distributed. What about the stellar surroundings of M13 versus M92? M13 has its two familiar flanking stars of 7th magnitude in the low-power view; M92 is in an area surprisingly lacking in stars even that bright. Experienced amateur astronomers with fairly large telescopes also know that there is an 11th-magnitude spiral galaxy only 0.5° from M13; M92 has no such close deep-sky company.

Many other differences among globular clusters are mentioned in the final section of this chapter, where we survey a number of the best of these objects. But first, let us take to a closer look at the most popular of globular clusters, the great M13.

A Beneficent Monster

M13 actually has close competition for the title of grandest globular cluster in the northern celestial hemisphere. Some people prefer M5. I've always had a special fondness for M3. But it's not just familiarity and convenience of location that make M13 so highly esteemed.

M13 *is* easy to find—if your skies are at least dark enough to allow you to locate the distinctive pattern of stars in Hercules that is called "the Keystone." If you need help latching onto which stars are the only moderately bright naked-eye stars of the Keystone, remember that it lies just over 60 percent of the way along the line from the brilliant Arcturus to the brilliant Vega. Once you've found the four-star Keystone (see the July all-sky map in the figure on page 16), you want to look about two-thirds of the way between Zeta and Eta Herculis, which form the western side of the Keystone. If you have fairly dark country skies, you will probably be able to glimpse a flicker or even a slightly larger spot of light in that position—you're seeing M13. Some books rate its brightness as magnitude 5.8, but expert observers tend to feel it is brighter (Steve O'Meara estimates it as 5.3).

Now, a 4-inch telescope serves to provide only a mild introduction to M13. You'll see a mystically glowing, roughly spherical mass of light. But with most telescopes of that size, even in fairly dark skies, you will probably not see much structure in the cluster or glimpse any individual members of it. Increase your aperture to 6 inches or 8 inches, however, and the mass starts to take on detail and personality, and you start glimpsing the pinpricks of the brightest stars in the outlying regions of the cluster. Use averted vision to see more of these specks of stellar gold dust and to start tracing out the lanes and curves of unresolved stars.

A 10-inch or larger telescope reveals M13 as an object of awe. The thought always strikes me that this is a beneficent monster. I think "monster" perhaps in part because M13 is especially noted for the arms, tendrils, tentacles, or "curvilinear branches" that extend far out from it. Can you sketch some of this fantastic complexity in a drawing? It may help you get your bearings in what may otherwise be a bewilderment, albeit a beautiful one. Will you see lanes of darkness more prominently in some parts of the cluster? Other globular clusters exhibit these lanes (those lanes are not always an optical illusion; sometimes they represent a real structure), but M13 is the most famous case. The best-known structure of dark lanes is a dark Y shape, which is centered southeast of the cluster's core. Not everyone can see it, perhaps partly because it is more noticeable with certain apertures of telescope at certain magnifications. Perhaps you will, like other people, find prominent dark lanes cutting elsewhere through M13's mighty blazing mass of concentrated stellar multitude.

The Other Great Globular Clusters

Although globular clusters appear at many distances and in many directions from the galactic center, we are most likely to see them looming here and there when we look toward the much larger fraction of our galaxy that lies in the direction of the galactic center. Because that center is located in Sagittarius, that constellation and neighboring ones contain more globulars than any other, and a lion's share of the globulars are visible in the part of the heavens we see on summer evenings. Among the greatest globular clusters properly visible to observers at midnorthern latitudes, however, one is in a spring constellation, five are in summer constellations, and two are in autumn constellations. But the spring and autumn clusters are situated in the late spring and the early autumn parts of the heavens. Winter has just one fairly easy globular cluster to see: M79 in Lepus the Hare, well south of Orion. But M79 is much dimmer than the greats we now will briefly profile.

Spring

The first mighty globular cluster of the year is M3. It shines within the bounds of Canes Venatici (the Hunting Dogs) in a location remote from any good guide stars. You'll find it a little less than halfway along a line drawn from Arcturus to Cor Caroli (Alpha Canes Venaticorum)—but that's a long line. M3 is worth the trouble. It seems to be more easily resolvable into stars than most of the great globular clusters (which is a good thing). Steve O'Meara finds M3 particularly colorful (with a peach-colored core!) and reminds viewers with sizable telescopes to look for dark patches near the center. M3 has the distinction of containing far more variable stars than any other globular cluster.

Spring evenings are when telescopic observers between about 40° N and 25° N can look down near their due south horizon to catch a low and therefore greatly dimmed glimpse of the greatest of all globular clusters—Omega Centauri. Only in the tropics or the southern hemisphere is Omega Centauri high enough to be appreciated for the immense magnitude 3.7 ball that it is.

Summer

We have already discussed M13 and M92 in Hercules. Down in the constellation Serpens Caput (the Serpent's Head) burns a globular cluster at least almost as glorious as M13—M5. This decidedly elliptical cluster lies only 20' from the lovely double-star 5 Serpentis. M4 is situated 1.3° west of a star—but what a star: Antares. M4 vies for the title of closest globular with the far southern cluster NGC 6397 (both may be slightly less than 7,000 light-years from

Earth—by comparison, M13 is about 20,000 to 25,000 light-years distant). M4 shows its closeness by its large apparent size, looseness, and resolvability of stars. It also possesses a mysterious bar of 11th-magnitude stars across its center that starts becoming visible in a 4-inch telescope but in very large telescopes helps make M4 one of the most magnificent of all deep-sky objects. If you don't have an enormous telescope, you can get great thrills by turning a 6-inch or 8-inch one on a globular cluster that is bigger and brighter and really may be more splendid than M13. But because this cluster is low, it is seen well by observers at midnorthern latitudes only on very clear, haze-free nights. I'm talking about the large, relatively close and loose M22, located just 2.3° northeast of Lambda Sagittarii (for more about it, see Sight 32).

Autumn

There are two great globulars of early autumn: M15 and M2. Strangely, they are located at almost exactly the same right ascension. M2 is about 13° due south of M15 and lies just south of the celestial equator in Aquarius. M15 is conveniently located only about 4° northwest of Enif (Epsilon Pegasi), the 2nd-magnitude nose of Pegasus. M2 and M15 are very similar in brightness and size and both are unusually elliptical. With an 8-inch or larger telescope, I have found M2 to be somewhat more colorful than M15.

A Bright Planetary Nebula

A diffuse nebula such as the Great Orion Nebula can provide the material from which new stars are born. But at the other end of a star's life, some stars may quietly cough out a cloud of gas and dust that is called a **planetary nebula**. These exhalations of a dying star have nothing to do with planets in the solar system of the star. The name is a throwback to the nineteenth century when observers noted the resemblance of many of these nebulae to the appearance of the planets Uranus and Neptune—small, round, fairly hard-edged, and blue-green. The eerie, intense blue-green light from a planetary nebula is produced when it is stimulated by the ultraviolet radiation from the hottest of all stars (180,000° F surface temperature): the dying white dwarf star that it surrounds. The color and often high-surface brightness of planetary nebulae combine to make them some of the most impressive of all deep-sky objects.

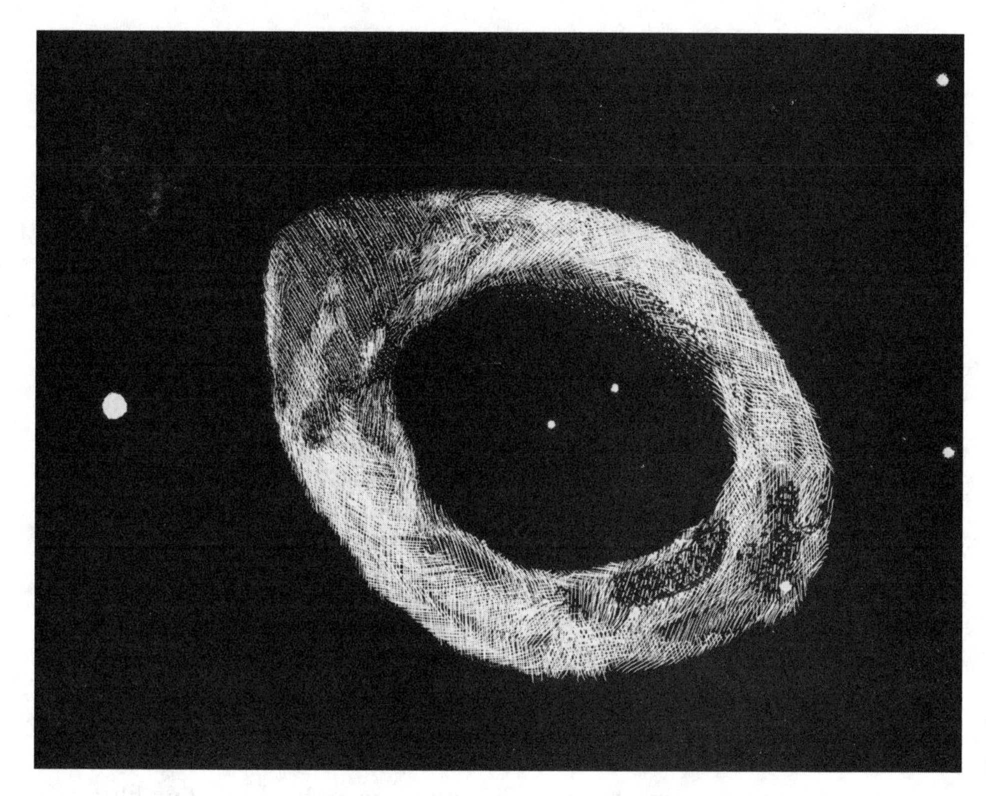

The Ring Nebula.

The Ring and the Dumbbell

Like other deep-sky objects, the most conspicuous planetary nebulae usually receive nicknames that refer to their shapes. Thus, the most famous planetary nebula is the Ring Nebula (M57) and the most brightly prominent is the Dumbbell Nebula (M27). (The largest and greatest in total brightness is the Helix Nebula [NGC 7293], but it is spread out over so large an area that its surface brightness is troublesomely low.)

The Ring Nebula is not only one of the very brightest planetary nebulae and striking in shape but is also located in an extremely convenient, conspicuous place. You'll find it on one side of the compact diamond shape that helps form Lyrae the Lyra: M57 lies slightly less than halfway along the line from Beta Lyrae to Gamma Lyrae. At very low magnification you may have to look hard to see it as a slightly fuzzy star, located within a triangle of rather dim telescopic stars. But as magnification and aperture of the telescope are increased, M57 turns first into a little glowing disk and then into a spooky, seemingly phosphorescent smoke ring. Some viewers have called it a cosmic doughnut. Maybe it is a deep-space life preserver. Whatever whimsical thoughts you may have, this object and its light are absolutely eerie and absolutely captivating.

Like most planetary nebulae, M57 bears magnification well. You may find that you can push a 6-inch telescope as high as 300× or an 8-inch as high as 400× on a good, steady night and still not lose, but possibly gain, detail in M57. Notice that one end of the somewhat tilted ring is less bright than the other. Even with a 4-inch telescope you shouldn't have trouble seeing a prominent star not far from M57. But that is not the *central star*, the ultrahot white dwarf that puffed out the nebula and that now energizes it with ultraviolet light. You would expect the central star to be within the hole of the ring. It is, but so is another, somewhat less dim star. The central star itself is apparently variable in brightness. But even if you were looking on a night when it was brighter than usual, it might only be 15th magnitude. That should be bright enough to see in a 10-inch telescope in really dark skies, but in reality even a considerably larger telescope may not show it to you. The reason? As even a medium-sized telescope on a good night can reveal, the hole of M57 is not completely dark—there is some background light within the hole of the Ring. This background radiance makes it harder to perceive the faint central star.

Another challenge with the Ring Nebula is trying to see more than one color in it. My eyes seem to be especially sensitive to color in low-light objects, so I may not be a representative case. But on some nights—presumably when the seeing was good—I have been able to detect the faint red, yellow, and blue in M57 with a 13-inch or even a 10-inch telescope. I've also detected several colors in our next showcase planetary nebula: M27, the Dumbbell Nebula.

Even though M27 is bigger and brighter than M57, there are two reasons that it is less popular than the Ring Nebula. First, its shape is not so regular or compelling to our sense of geometry as M57's. Second, it is not so conveniently located in relation to prominent guide stars. M27 is located in the Summer Triangle but in the inconspicuous constellation Vulpecula the Little Fox. With a Dobsonian mount, a trick I always use to locate the Dumbbell is to trace the pattern of Sagitta the Arrow along the shaft to its point (from Delta Sagittae to Gamma Sagittae), then make a right angle turn and move the telescope about the same distance as the length of the Sagitta arrow. At low power, this should get you close enough to pick up the big, bright M27, located just 0.5° south of a magnitude 5.7 star. If you have an equatorial telescope, all you have to do is move it 3.3° due north of Gamma Sagittae. M27 itself shines with a total brightness of magnitude 7.6—greater than any other planetary save for the giant Helix Nebula (NGC 7293 in Aquarius). But the

Helix is so big—almost half as big as the Moon—that its brightness is spread out over a large area. This makes it a ghostly object that is not easy to see except at low magnification in a dark sky. By contrast, the Dumbbell is big—8' by 5'—but not big enough to minimize its surface brightness. M27 is definitely the most prominent of all planetary nebulae.

But does it really look like a dumbbell? Yes, in fairly small telescopes it shows two lobes with a thinner region between and does look somewhat like a dumbbell. In medium-sized telescopes, however, the appearance is really more like an hourglass or an eaten apple that has only the core left. As larger—10-inch and more—telescopes are used in dark skies, the apple starts filling back in! Just as there is light within the hole of the Ring Nebula, so, too, here there is faint radiance that becomes more prominent with larger telescopic apertures.

There is a fairly bright star that is perched beautifully near the tip of one of the projections that stick out on both ends of the "apple core." But this is not the central star of M27. The central star shines at only about magnitude 13.9 and is very hard to see against the nebula's bright background.

Other Bright Planetary Nebulae

The Ring Nebula, the Dumbbell Nebula, and the Helix Nebula all look unusually large and possess great total apparent brightness because they are closer to us than most other planetary nebulae (they are about 1,800, 1,000, and 500 light-years away, respectively). Most others visible in amateur telescopes are several thousand light-years away and appear as small but intense disks of light. The closest planetary nebulae are near enough for us to either see (faintly) or at least photograph (easily) a range of different wavelengths of light (different colors) in them. But the rest of the planetary nebulae are far enough away that the radiance we see from them is entirely dominated by a very narrow range of blue-green wavelength.

Even the nicknames of the planetary nebulae are fascinating. Who wouldn't want to look for objects called the Ghost of Jupiter, the Saturn Nebula, the Eight-Burst Nebula, the Eskmo Nebula, and the Owl Nebula? By the way, one of the nicknames deserves explanation because it refers to a very interesting observational effect: the Blinking Planetary. This object, NGC 6826 in Cygnus, has a central star almost as bright as the nebula itself. If you stare directly at the star through a medium-sized telescope, the nebula seems to fade out. But when you switch to averted vision, the nebula appears and drowns out the star. Thus, an observer can quickly blink the nebula on and off.

Sight 50 A Bright and Structured Galaxy

What is our last sight in this book? We've seen M31, the Great Andromeda Galaxy, back in Sight 33. We've scanned through the bewilderingly rich Realm of the Galaxies back in Sight 34. But there are a number of other bright and structured galaxies that invite our close inspection of their distant wonder.

The Galaxies of Fall

Where the dusty band of the Milky Way's equatorial plane passes—across the skies of summer and winter—few galaxies can be seen. Spring and fall are the great seasons for galaxies.

Fall is, of course, dominated by the mighty Andromeda Galaxy. Unless you live where fall is spring—that is, where the autumn of Earth's northern hemisphere, the months of September, October, and November, are spring. In the

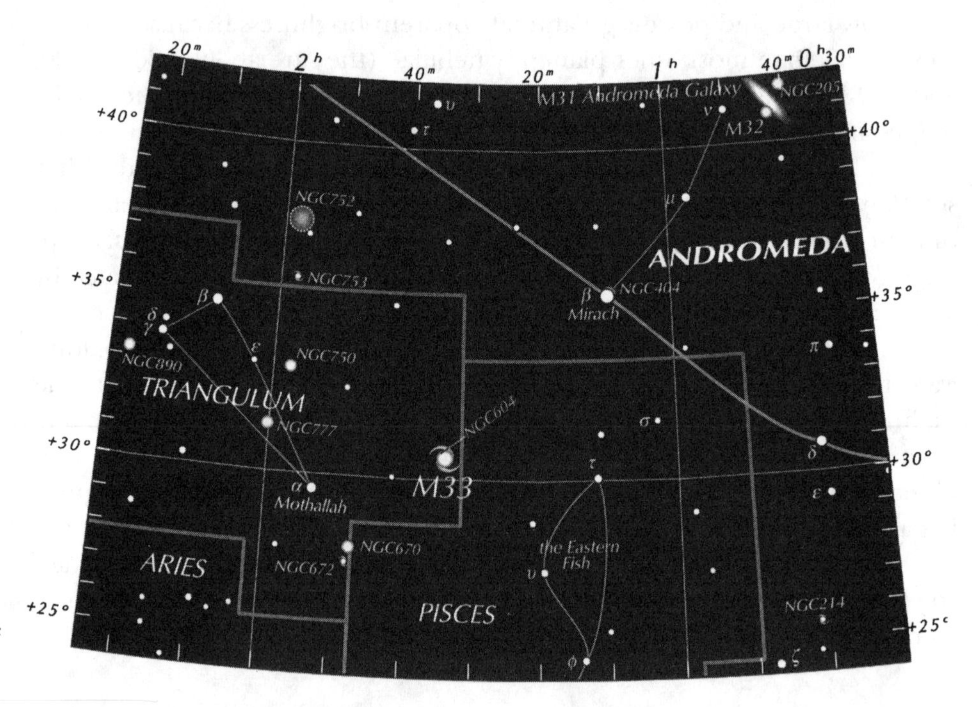

The locations of M33 and M31.

southern hemipshere lands such as Australia, New Zealand, South Africa, and Argentina and other South American countries, this is the time of year when the Milky Way's most prominent satellite galaxies soar high. They are the Large Magellanic Cloud and the Small Magellanic Cloud, and they do look like huge pieces of the Milky Way band that have gotten loose and blown away to their own section of the heavens.

Back in the northern heavens, M31 has a counterpart that lies about as far southeast of 2nd-magnitude Beta Andromedae as M31 lies northwest of that star. Just a few degrees from the star at the apex of Triangulum the Triangle floats a big ghost just barely visible to the trained naked eye in very dark skies: it is M33, the Pinwheel Galaxy.

M33 is notorious for sometimes being easier to see in a finderscope than in the eyepiece of a 10-inch telescope. The reason is that it is a virtually face-on spiral galaxy—and so (relatively) close that its brightness is spread out over a huge area. M33 measures as much as 1° across and your view of it through a telescope should be at low magnification in dark, clear skies. You should see the central area quite easily. Then start trying to behold outer parts and make sense of their structure with patient use of averted vision. The brightest of the outer clumps of M33 is the surprisingly prominent nebula NGC 604. It stands out all by itself in your telescope like a little orphan patch of light. The true size of this seemingly little puff is probably more than a

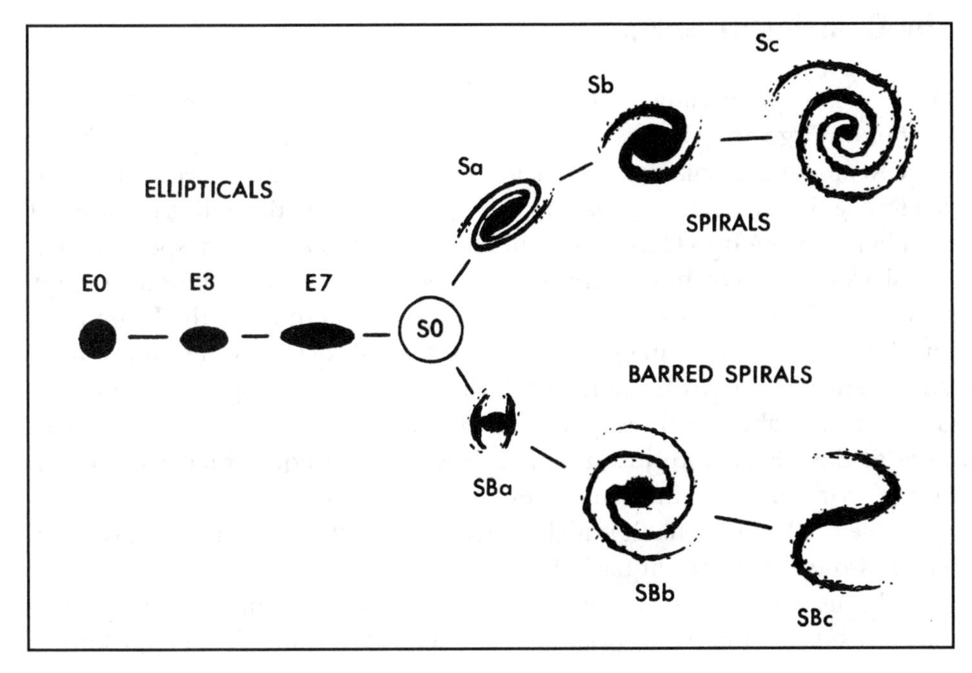

A diagram of the types of structured galaxies (no suggestion of an evolutionary scheme is intended).

thousand times that of the Great Orion Nebula! After all, we are seeing NGC 206 at M33's distance of, like M31, over 2 million light-years. M33 might even be a little farther than M31.

Startlingly different from M33 is a galaxy only 1° southeast of Delta Ceti, the star at the base of the head of Cetus the Whale. This is M77. It shines at only magnitude 8.9, compared to M33's total magnitude of about 6.3 (or even brighter). But M77 is compact, and it is most notable for having an intense center. This kind of center is the trademark of a Seyfert galaxy, whose center seems to be a mild version of those powerhouses of the universe, the **quasars**. M77 is the Seyfert galaxy of greatest apparent brightness.

Whereas M77 is essentially right on the celestial equator, we have to plunge 25° south from the equator to find another fine galaxy of fall. It is about 7° almost due south of the 2nd-magnitude tail of Cetus (Beta Ceti, also called Diphda and Deneb Kaitos). This galaxy, which is just over the border from Cetus in the constellation Sculptor, is NGC 253. It is a lovely spiral, magnitude 7.1 and 22' by 6'. With a 10-inch or 12-inch telescope, it begins to look almost like a much smaller and dimmer version of M31—though, of course, with many different details from the Great Andromeda Galaxy. On a clear, dark night, viewers at midnorthern latitudes can catch this galaxy at its highest and enjoy its details of beauty with fairly big telescopes.

The Galaxies of Spring

Of course, we've already explored spring's greatest galaxy wonderland: the regions of Virgo and Coma Berenices dominated by the Virgo Galaxy Cluster ("the Realm of the Galaxies"). Back in that chapter (Sight 34), we also looked at a few fascinating galaxies that are in Virgo and Coma Berenices but are not members of the Virgo Cluster: needle-thin NGC 4565 and dust-spotted M64, the Black-eye Galaxy. In that chapter, however, I merely mentioned a mighty galaxy that I said may (or may not) be a near-side member of the Virgo Cluster. I'm referring to the magnitude 8.3, 9'-by-4'-galaxy that shines in southern Virgo near the Corvus border: M104, the Sombrero Galaxy. The Sombrero Galaxy is arguably the most dramatic of all galaxies with prominent dark lanes of dust. Even with just a 6-inch telescope this equatorial strip of dust begins to become visible. In larger telescopes, M104 really does start to resemble the Mexican hat it is nicknamed after. In this case, the brim is delineated with the vast strip of dark dust.

Did I just say that the Sombrero Galaxy was the galaxy made most spectacular by its dust lane? Then what about NGC 5128—the object better known as Centaurus A? Centaurus A is the result of colliding galaxies and appears as a roundish patch with an amazing bisecting dark lane—if one can call this strange feature a lane. The reason it is not known better to most amateur astronomers is that it is a quite southerly object—it never gets high enough to see properly for observers around 40° N.

The spiral galaxies that show dark streaks are usually ones that we see edge-on or at low angle to us. In contrast are the face-on galaxies of spring, which like M33, can suffer from a problem with surface brightness. The most intense of these is M51—the wonderful Whirlpool Galaxy just under the Big Dipper's handle in Canes Venatici (the Hunting Dogs). An 8-inch (or slightly smaller) telescope suffices to start showing glimpses of the spiral structure of M51. Under very dark skies such a telescope might also gather just enough light to show what looks like a connecting bridge of light between M51 and a strange, smaller, glowing mass. That mass is another galaxy that is not actually attached to a spiral arm of M51 but is in reality a little more distant than the big galaxy. Whatever the physical explanation for their closeness, the appearance of M51 and this companion is truly marvelous.

Face-on spiral galaxies of spring that have lower surface brightness than M51 but are well worth examination (especially with larger telescopes) are M101 (north of the handle of the Big Dipper) and M83, far south of the other two in Hydra.

You will want to visit the Leo trios—the galaxy triads M95, M96, and M105 and M65, M66, and NGC 3628. But much brighter and truly spectacular is the most impressive telescopic pair of galaxies in all the heavens: M81 and M82. This galaxy odd-couple shimmers in the far north—declination +69°—in Ursa Major. M81 is a magnitude 7.0 spiral of great beauty and size (26' by 14'). But just 38' from it shines the peculiar cigar-shaped M82, at magnitude 8.4 and 11' by 5'. Even small telescopes can show a pretty good view of these two galaxies, for they are central members of a gathering not far beyond our Local Group of galaxies. The larger the aperture you use, the more detail you see in both galaxies. In large amateur telescopes, the amount of mottling and other fine structure visible in M82 are amazing. There are two other, much dimmer galaxies in the vicinity of M81 and M82. But it is this pair, made all the more dramatic by their contrast with each other, that capture the imagination. What caused M82 to look like it does? Was it (as astronomers seem to think) an interaction with M81 as the two passed each other? Whatever the answers may be, you'll be pulled far outside of yourself into the deepest midnight of space by this eerie vision. You'll feel you are truly out there watching from not far away the handsome and orderly magnificence of M81 being accompanied by the phantom spaceship or hoary phosphorescent cosmic whale, or unknown species of celestial leviathan that is M82.

APPENDIX A

Total Solar Eclipses, 2008–2024

In the following table, usually only parts of each country or ocean mentioned see a total eclipse. Discussion of annular and partial solar eclipses can be found in Sight 37.

Date	Maximum Duration (in minutes and seconds)	Regions of Visibility
2008 August 1	2:27	North Canada, Greenland, Arctic Ocean, Russia, Mongolia, China
2009 July 22	6:39	India, Nepal, Bhutan, Burma, China, Pacific Ocean
2010 July 11	5:20	South Pacific Ocean, Easter Island, Chile, Argentina
2012 November 13	4:02	Northern Australia, South Pacific Ocean
2013 November 3	1:40	Annular-total: total in Atlantic and Gabon to Kenya, annular in Ethiopia
2015 March 20	2:47	North Atlantic Ocean, Faeroe Islands, Arctic Ocean, Svalbard
2016 March 9	4:10	Indonesia, Pacific Ocean
2017 August 21	2:40	Pacific Ocean, United States (Oregon, Idaho, Wyoming, Nebraska, Missouri, Illinois, Kentucky, Tennessee, North Carolina, South Carolina), Atlantic Ocean
2019 July 2	4:33	South Pacific Ocean, Chile, Argentina
2020 December 14	2:10	Pacific Ocean, Chile, Argentina, South Atlantic Ocean
2021 December 4	1:55	Antarctica
2023 April 20	1:16	Annular-total: total in southern Indian Ocean, western Australia, Indonesia, Pacific Ocean
2024 April 8	4:28	Pacific Ocean, Mexico, United States (Texas, Oklahoma, Arkansas, Missouri, Kentucky, Illinois, Indiana, Ohio, Pennsylvania, New York, Vermont, New Hampshire, Maine), southeastern Canada, Atlantic Ocean

APPENDIX B

Major Meteor Showers

Shower	Peak[a]	Above Q-Max[b]	Number/Hour[c]	Time[d]	Radiant Location[e]
Quadrantids	January 4	January 4[f]	40[g]	6 A.M.	15h 28m, +50°
Lyrids	April 22	April 21–23	15	12 A.M.	18h 4m, +34°
Eta Aquarids	May 5	May 1–10	10	4 A.M.	22h 30m, –2°
Delta Aquarids	July 29	July 19–August 8	25	2 A.M.	22h 30m, 0° and 22h 40m, –16°
Perseids	August 12	August 9–14	50	4 A.M.	3h 4m, +58°
Orionids	October 21	October 20–25	25	4 A.M.	6h 12.5m, +13.5° and 6h 25m, +19.5°
Taurids	November 3	October 20–November 30?	10	12 A.M.	3h 32m, +14° and 4h 16m, +22°
Leonids	November 18	November 16–20	5[h]	5 A.M.	10h 8m, +22°
Geminids	December 14	December 12–15	50	2 A.M.	7h 28m, +32°

[a]May vary by a day or so because of leap years.
[b]The period when the shower's rates are above quarter the maximum strength.
[c]The number of meteors per hour typcially seen at the peak under dark, clear skies.
[d]The time (standard or daylight saving time) of night when the shower's radiant is highest outside of twilight.
[e]Where the shower meteors appear to be streaking away from, given in right ascension and declination (see discussion of these terms in Basic Information for Astronomical Observers in the front of this book).
[f]Only about 14 hours above quarter-maximum strength.
[g]Much higher rates in some years.
[h]Tremendously higher rates in some years.

APPENDIX C

Total and Partial Lunar Eclipses, 2007–2017

The following table does not list the penumbral lunar eclipses (which are of much less interest) in the period. The eclipse of October 8, 2014, features an incredibly rare event for observers in central and northern Alaska and Canada: an occultation of Uranus during totality. The eclipse of April 4, 2015, features the shortest duration of lunar totality—12 minutes—since 1856. There are no total or partial eclipses in 2016.

Date	Type	Time[a]	Regions of Visibility	Magnitude[b]	Duration (in minutes)
2007 March 3	Total	23:21	Eastern Americas, Africa, Europe, Central Asia	1.24	74
2007 August 28	Total	10:38	Eastern Asia, Australia, Pacific, western North America	1.48	90
2008 February 21	Total	3:26	Americas, Europe, Africa	1.11	50
2008 August 16	Partial	21:10	Eastern South America, Africa, Europe, Central Asia	0.81	
2009 December 31	Partial	19:23	Europe, Africa, Asia, western Australia	0.08	
2010 June 26	Partial	11:39	Australia, Indonesia, Pacific, western Americas	0.54	
2010 December 21	Total	8:17	Northeast Asia, Pacific, Americas	1.26	72
2011 June 15	Total	20:13	Africa, South Asia, Australia	1.71	100
2011 December 10	Total	14:32	Central and East Asia, Australia, Alaska, northern Canada	1.11	50
2012 June 4	Partial	11:04	Central and eastern Australia, Pacific, western North America	0.38	
2013 April 25	Partial	20:08	Central Africa and Asia	0.02	
2014 April 15	Total	7:46	Americas	1.30	78
2014 October 8	Total	10:55	East Asia, eastern Australia, Pacific, central and western North America	1.17	58
2015 April 4	Total	12:01	Eastern Australia, Pacific, western North America	1.01	12
2015 September 28	Total	2:48	Eastern North America, South America, western Europe, western Africa	1.29	72
2017 August 7	Partial	18:21	Central and East Asia, Australia	0.25	

[a]Universal Time of greatest eclipse.
[b]How much of the Moon's diameter is covered by umbra at greatest eclipse.

APPENDIX D

The Brightest Stars

Name	Greek Letter Designation	Apparent Magnitude	Absolute Magnitude	Distance (in light-years)
Sirius	Alpha Canis Majoris	–1.44	1.5	9
Canopus	Alpha Carinae	–0.62	–5.4	313
Alpha Centauri	Alpha Centauri	–0.28[a]	4.2	4
Arcturus	Alpha Bootis	–0.05	–0.6	37
Vega	Alpha Lyrae	0.03	0.6	25
Capella	Alpha Aurigae	0.08	–0.8	42
Rigel	Beta Orionis	0.18	–6.6	773
Procyon	Alpha Canis Minoris	0.40	2.8	11
Achernar	Alpha Eridani	0.45	–2.9	144
Betelgeuse	Alpha Orionis	0.45[b]	–5.0	522[c]
Beta Centauri	Beta Centauri	0.61	–4.4	335
Alpha Crucis	Alpha Crucis	0.75[a]	–4.4	340
Altair	Alpha Aquilae	0.76	2.1	17
Aldebaran	Alpha Tauri	0.87[b]	–0.8	65
Spica	Alpha Virginis	0.98[b]	–3.6	262
Antares	Alpha Scorpii	1.06[b]	–5.8	604[c]
Pollux	Beta Geminorum	1.16	1.1	34
Fomalhaut	Alpha Piscis Austrini	1.17	1.6	25
Beta Crucis	Beta Crucis	1.25	–4.0	352
Deneb	Alpha Cygni	1.25	–7.5	1467[c]
Regulus	Alpha Leonis	1.36	–0.6	77

[a]Combined magnitude of visual double.
[b]Variable star.
[c]Distance (and therefore absolute magnitude) significantly uncertain.

Additonal names for some of these stars include Rigilkent for Alpha Centauri, Hadar and Agena for Beta Centauri, Acrux for Alpha Crucis, and Becrux or Mimosa for Beta Crucis.

APPENDIX E
Transits of Venus and Mercury

Transits of Venus, 1874–2125

1874 December 9

1882 December 6

2004 June 8

2012 June 6

2117 December 11

2125 December 8

Transits of Mercury, 2003–2052

2003 May 7

2006 November 8

2016 May 9

2019 November 11

2032 November 13

2039 November 7

2049 May 7

2052 November 9

GLOSSARY

absolute magnitude The magnitude of brightness a star would have if seen at a standard distance of 10 parsecs (about 32.6 light-years).

altazimuth system System for indicating positions in the sky using altitude and azimuth as vertical and horizontal measure.

altitude Apparent angular height in the sky (vertical measurement in the altazimuth system).

apparent magnitude The magnitude of brightness a star appears to have in the sky.

asterism A pattern of stars in the sky that is not an official constellation.

asteroid A rocky body between several hundred yards and several hundred miles wide (in most but not all cases located between the orbits of Mars and Jupiter).

atmospheric extinction Dimming of the light of celestial objects because of absorption and scattering by Earth's atmosphere.

aurora Patterns of radiance (usually moving and fluctuating) produced when atomic particles from the Sun are energized in Earth's magnetic field and channeled to collide with upper-atmosphere gases in regions surrounding Earth's magnetic poles. Also known as "the Northern Lights" or "aurora borealis" ("aurora australis" in the southern hemisphere).

averted vision A technique of looking slightly to the side of a faint object to increase its visibility by allowing its light to fall on the parts of the eye's retina most sensitive to light.

azimuth Horizontal measure around the sky in the altazimuth system.

binary star A double-star system in which the members are believed to be orbiting around each other (that is, around a common center of gravity).

black hole An object, thought to be the result of a massive star's collapse, whose gravity has become so strong as to prevent even light (and other electromagnetic radiations) from escaping it.

blue giant A massive, large, and extremely luminous star with a very high surface temperature that causes it to appear bluish-white to the eye.

celestial objects Bodies in outer space beyond Earth's atmosphere.

celestial sphere The imaginary sphere surrounding Earth whose inner surface is all the sky both above and below one's horizon.

Cepheid A type of pulsating variable star whose precisely regular brightness variations can be employed to estimate the star's distance by use of the period-luminosity relation (if you know the period of the Cepheid's brightness variations, you know its luminosity—true brightness—and can therefore compare this to its apparent brightness to determine distance).

circumpolar Close enough to one of the celestial poles so as to never rise or set but rather circle around the pole above the horizon.

comet A mostly icy body, tremendously less massive than the planets, that produces a cloud of dust and/or gas (coma) when it is heated sufficiently by the Sun (or, less commonly, by other forces).

conjunction Strictly speaking, the arrangement when one celestial object moves to have the same right ascension or ecliptic longitude (passes due north or due south of a second object in a celestial coordinate system) as another. More loosely, any temporary pairing or gathering of celestial objects that is considered close.

constellation An official pattern of stars or, more strictly, the officially demarcated section of sky in which that pattern lies.

crescent Phase in which a world's hemisphere that is facing us appears less than half lit.

culminate Reach the north-south meridian of the sky, typically achieving the highest object's point above the horizon.

dark adaptation The increase in the sensitivity of our eyes to dim light that occurs when they are kept away from bright light for a while.

dark nebula A nebula that does not shine by either emitted or reflected light and is therefore visible only in silhouette against a more distant bright nebula or starry background.

declination North-south measure in the equatorial system of celestial coordinates, corresponding to latitude on Earth.

deep-sky object An object beyond our solar system, though the term is usually not applied to individual stars or double and multiple star systems but to star clusters, nebulae, and galaxies.

diffuse nebula A luminous nebula that shines from reflecting the light of nearby stars (reflection nebula) or is heated enough by very hot stars to glow on its own (emission nebula).

diurnal motion The apparent motion of celestial objects caused by the Earth's rotation.

double star A star that, upon closer or more sophisticated examination, turns out to consist of two or more component stars.

eclipse The hiding or dimming of one object by another object or by another object's shadow.

eclipsing binary A type of variable star in which one component star of a double star eclipses the other, or both alternately eclipse each other, causing the variations in brightness.

ecliptic The apparent path of the Sun through the zodiac constellations, which is really the projection of Earth's orbit in the sky.

equatorial system A system for indicating positions in the heavens using right ascension (corresponding to longitude on Earth) and declination (corresponding to latitude on Earth).

fireball A meteor brighter than Venus.

full-cutoff fixture A light fixture that emits light entirely below the horizontal, eliminating directly produced skyglow and reducing light pollution.

galactic cluster See **open cluster**.

galaxy An immense congregation of typically billions of stars forming a system of spiral, elliptical, or irregular shape.

gibbous Phase in which the hemipshere of a world facing us is more than half lit but less than fully lit.

globular cluster A kind of star cluster consisting of tens of thousands up to a few million stars arranged in a roughly spherical shape.

greatest elongation The maximum angular separation of an inferior planet (Mercury or Venus) from the Sun.

horizon The boundary line between sky and land or sea on Earth.

inferior conjunction An inferior planet's passage of the Sun on the near side of its orbit to Earth (see **superior conjunction**).

inferior planet A planet closer to the Sun in space than the Earth is (contrast **superior planet**).

light pollution Excessive or misdirected artificial outdoor lighting.

light-year The distance that light, the fastest thing in the universe, travels in the course of one year.

limb The edge of the Moon or other celestial body.

long-period variable A major kind of variable star in which the period of brightness variations is months or years long and the range of the variations are typically great, with both often being irregular or only semiregular.

luminosity The true brightness of a star, independent of its distance from us, measured in units of the Sun's true brightness (for instance, a star twice as luminous as the Sun has a lumiosity of 2).

magnitude A measure of brightness in astronomy in which an object 100 times brighter than another is exactly 5 magnitudes brighter. The brighter the object, the lower the magnitude figure (e.g., a 1st-magnitude star is brighter than a 2nd-magnitude star), with negative magnitudes for the very brightest objects of all.

meridian The line in the sky that passes from the due south horizon to the zenith onward to the due north horizon.

Messier objects (also known as M-objects) More than 100 deep-sky objects cataloged in the eighteenth century by the French astronomer Charles Messier.

meteor The streak of light seen when a piece of space rock or iron enters Earth's atmosphere at high speed and becomes luminous from its friction.

meteorite A piece of space rock or iron that survives its trip through the atmosphere as a meteor and reaches the ground.

meteoroid A rocky or metallic object smaller than an asteroid (dust-sized to no more than a few hundred yards across) that would become a meteor if it entered Earth's atmosphere and a meteorite if it reached Earth's surface.

meteor shower An increased number of meteors, all appearing to diverge from the direction of a single area among the constellations.

meteor storm An extremely intense meteor shower in which hundreds or even thousands of meteors per hour may be observed.

meteor train The trail of ionization that glows in the sky after some meteors have disappeared.

Milky Way The spiral galaxy that we live in and also the night sky's band of strongest glow from the combined light of innumerable distant stars in the galaxy's equatorial plane.

moon A rocky or icy object that circles a planet. Also known as a (natural) satellite.

multiple star A star system consisting of more than two stars (although double star is often used as the umbrella term for systems of two, three, four, or more stars).

nebula A vast cloud of dust and gas in interstellar space. Different types include diffuse nebulae (which include emission nebulae and reflection nebulae), planetary nebulae, and dark nebulae.

neutron star The collapsed, ultradense core of a star left after a supernova formed from an original star not massive enough to collapse all the way into becoming a black hole.

Northern Lights See **aurora**.

nova An exploding (and therefore briefly brightened) star that loses a small fraction of its mass in the outburst, which may often arise from interactions between stars in double-star systems.

occultation The hiding of one celestial object by another that is usually much larger (in a few cases, where the two bodies are not greatly different in apparent size, a hiding can be called an eclipse).

open cluster (also called galactic cluster) This major kind of star cluster consists of a usually irregular shape and includes typically dozens or a few hundred stars.

opposition Position of a superior planet when it is directly opposite the Sun in the sky—thus rising around sunset, highest around the middle of the night, and setting around sunrise (this is also the position at which a superior planet is usually closest to us and appears biggest and brightest).

optical double A double star in which the two components are not truly related, one object being much farther away and just happening to lie on nearly the same line of sight as seen from Earth.

PA See **position angle**.

parallax The change in a star's position caused by our change in viewpoint (usually our change from one side of Earth's orbit to the other).

parsec A parallax-second, the distance at which the view from opposite sides of Earth's orbit would cause an object to have an apparent position change (a parallax) of 1 arc-second (1 parsec is equal to about 3.26 light-years).

planet A relatively massive world in direct orbit around a star.

planetary nebula A cloud of gas and dust cast off by a small, hot, dying white dwarf star. (The name comes from the passing resemblance of some of these blue or green nebulae to the planets Uranus and Neptune as seen in the telescope.)

position angle (PA) The direction angle of a companion star in relation to its primary in a double-star system.

precession A slight wobble in Earth's rotational axis, caused by the gravitational pull of the other solar system bodies and resulting in slow changes of the direction of the north celestial pole and other positions in the heavens.

proper motion The motion of a star relative to the Sun as projected on the celestial sphere—in other words, the change in a star's position on the celestial sphere produced by the component of its space velocity (motion through space) that is transverse (neither toward nor away from us).

pulsar A type of neutron star oriented toward Earth in such a way that we get to observe pulses of light and other types of eletromagnetic radiation released from gaps in the star's magnetic field near its poles.

quasar An incredibly powerful source of light and other types of electromagnetic radiation that may be a kind of intense core of a galaxy.

RA See **right ascension**.

radiant The region in the sky from which the meteors of a shower all appear to diverge.

red dwarf A small star far less massive than the sun that radiates relatively little light and mostly in the red because of its comparatively low surface temperature.

red giant A huge and fairly massive star of extremely low density that radiates mostly in the red because of its relatively low surface temperature.

revolution The orbiting of one celestial body around another (the Earth's revolution period around the Sun is one year).

right ascension (RA) West-east measure in the equatorial system of celestial coordinates, corresponding to longitude on Earth (though expressed somewhat differently than longitude on Earth: in hours of right ascension from 0 to 24).

rotation The spinning of a celestial object (the Earth's rotation period is 24 hours).

satellite Any celestial body that orbits another, but usually this term is confined to a body that orbits a planet. An artificial satellite is one launched by humans to orbit Earth or another world; a natural satellite (composed of rock and/or ice) is more popularly known as a moon.

seeing Sharpness of astronomical images as a function of turbulence in Earth's atmosphere.

sidereal time Time measured by the passage of the stars around the sky, without reference to the Sun (the sidereal day is about 4 minutes shorter than the solar day).

skyglow The component of light pollution (excessive and misdirected artificial outdoor lighting) that goes up into the sky. It is visible for dozens of miles around even just fairly large cities.

SNR See **supernova remnant**.

solar system The whole collection of planets, moons, asteroids, comets, and meteoroids orbiting the Sun, or another star, under the Sun's or other star's gravitational influence.

spectral class The subcategory of spectral type in which a star belongs—for instance, O3, G2, M5—in which the letter denotes the type and the number indicates the class.

spectral type The category—ranging from O (the hottest normal stars) to M (the coolest normal stars), with some alternative unusual types—into which a star is placed according to the appearance of its light's chemical spectrum.

sporadic meteor A meteor not traceable to any known meteor shower.

star A massive self-luminous ball of gas producing energy by nuclear fusion in a dense core (also called a sun or, in the case of Earth's star, the Sun).

star cluster A grouping of a few to several million stars traveling through space relatively close together but not closely enough to be considered a multiple star.

superior conjunction A superior planet's passage of the Sun on the far side of its orbit to Earth (contrast **inferior conjunction**).

superior planet A planet farther from the Sun in space than the Earth is (contrast **inferior planet**).

supernova A powerful kind of star explosion and brightening in which a large part of the star's mass is lost and its core may become a neutron star or black hole.

supernova remnant (SNR) The cloud of material ejected by a supernova, sometimes visible for many thousands of years after the explosion.

transparency The degree to which Earth's atmosphere is capable of letting celestial light pass through it (how clear of dust and water vapor is the atmosphere over you tonight?).

Universal Time (UT) Time system for dating astronomical events, corresponding to the local time at Greenwich, England, on the 0° meridian of longitude on Earth (to obtain Universal Time from your current local standard time, subtract 5 hours from eastern standard time, 4 hours from central standard time, and so on).

variable star A star that for one of a number of possible reasons undergoes changes in its brightness.

white dwarf A hot, extremely small but fairly massive (and therefore very dense) star that represents the last luminous stage in the lives of many stars.

zenith The overhead point in the sky.

zodiac The circle of constellations through which the Sun passes during the course of the year.

SOURCES

Books

Burnham, Robert, Jr. *Burnham's Celestial Handbook.* New York: Dover Publications, 1978. The classic and nearly comprehensive guide to deep-sky objects visible through telescopes up to about 10-inch aperture. Some of the information has become outdated, but the combination of vivid descriptions, useful tables, and inspired writing about lore and science has not been surpassed.

Harrington, Phil. *Eclipse! The What, Where, When, Why, and How Guide to Watching Solar and Lunar Eclipses.* New York: John Wiley & Sons, 1997. A very thorough and engaging book about both lunar and solar eclipses, with details about individual eclipses through 2017.

———. *Touring the Universe through Binoculars.* New York: John Wiley & Sons, 1990. An excellent constellation-by-constellation tour of binocular sights, along with information about selection and care of binoculars.

Houston, Walter Scott. *Deep-Sky Wonders.* Cambridge, Mass.: Sky Publishing Corporation, 1999. A collection of essays from the great Walter Scott Houston's seminal and enduring column about deep-sky objects.

Littmann, Mark, Ken Wilcox, and Fred Espenak. *Totality: Eclipses of the Sun.* New York: Oxford University Press, 1999. The most complete book on the history and observation of solar eclipses ever written.

Mullaney, James. *Celestial Harvest.* New York: Dover Publications, 2002. One of the world's most veteran deep-sky observers offers wonderful descriptions, along with data, about more than 300 of the finest deep-sky objects.

———. *Double and Multiple Stars and How to Observe Them.* London: Springer, 2005. An excellent, spirited guide by probably the world's most experienced double-star observer.

O'Meara, Stephen James. *The Messier Objects.* Cambridge, Mass. and Cambridge, England: Sky Publishing Corporation, 1998. The great observer O'Meara discusses each of the 110 Messier objects in detailed and often poetic descriptions along with a photograph, sketch, and more for each object.

Ottewell, Guy. *The Astronomical Companion.* Greenville, S.C.: Ottewell, 2000. Ottewell's seventy-two-page atlas-sized guide to astronomical topics, including eclipses, lunar phases and motions, types of stars, and stellar evolution, which are year-independent (see his annual "Astronomical Calendar" for astronomical phenomena that are year-dependent); includes dozens of huge and unique diagrams.

———. *The Under-standing of Eclipses.* Universal Workshop, 1990. The most inspired and original book on eclipses ever written, with unique diagrams.

Schaaf, Fred. *40 Nights to Knowing the Sky.* New York: Henry Holt and Company, 1998. A comprehensive introduction to the workings and appearances of the Moon, the Sun, planets, stars, and deep-sky objects on the celestial sphere and in the universe.

———. *Seeing the Sky.* New York: John Wiley & Sons, 1990. *Seeing the Solar System.* New York: John Wiley & Sons, 1991. *Seeing the Deep Sky.* New York: John Wiley & Sons, 1992. A trilogy of books of observing projects focusing on naked-eye observations, telescopic observations of solar system objects, and telescopic observations of objects beyond the solar system.

Wood, Charles A. *The Modern Moon: A Personal View.* Cambridge, Mass.: Sky Publishing Corp., 2005. A superb book by Sky & Telescope's Moon columnist, this authoritative work not only tells readers what features are visually compelling to view on the Moon but also provides the fascinating scientific explanations of their nature.

Atlases, Sky Simulation Software, Planispheres

Star atlases range from those displaying only stars down to 5th or 6th magnitude (a few thousand stars) to *The Millennium Star Atlas* (well over 1 million stars, all stars brighter than magnitude 11.0). A fine selection of these atlases are described in the Sky Publishing Corporation catalog (see the Web address for "Sky & Telescope" magazine in "Online" sources on page 272).

An amazingly concise and informative little atlas/handbook is E. Karkoschka's *The Observer's Sky Atlas* 2nd ed. New York: Springer, 1999.

The best lunar atlas is surely Antonin Rukl's *Atlas of the Moon.* Cambridge, Mass.: Sky Publishing Corp., 2005.

Various sky simulation software programs offer access to maps of even dimmer stars. Descriptions of many of these can also be found in the Sky Publishing catalog and elsewhere (catalogs of the Astronomical Society of the Pacific and Orion Telescope and the Binocular Center, for instance).

If you need star-finding information that you can carry around and out to your observing site easily, try an old-fashioned planisphere, a handheld rotating wheel of cardboard or plastic that displays the starry sky for any time of any night. The best all-around planisphere is probably the *Precision Planet and Star Locator*, produced by David Kennedal and distributed in North America by Sky Publishing Corporation.

Periodicals, Almanacs, Annual Guides

Astronomical Calendar. Eighty-two atlas-sized pages filled with original diagrams and rich text about the year's celestial events (includes several sections and month-by-month "Observers' Highlights" by the author of the book you are holding). By Guy Ottewell at Universal Workshop (www.universalworkshop).

Astronomy (www.astronomy.com). One of the two largest and most popular astronomy magazines (Sky & Telescope being the other).

Mercury. A fine magazine published by the Astronomical Society of the Pacific (www.astrosociety.org).

Observer's Handbook (www.rasc.ca/handbook). An excellent annual guide with very extensive technical information for observers written by a variety of experts and published by the Royal Astronomical Society of Canada.

Sky & Telescope (www.SkyTonight.com). The most enduring of astronomy magazines. It includes planets and stars columns (plus other pieces) by the author of this book.

Sky Calendar. Abrams Planetarium, Michigan State University, East Lansing, MI 48824 and (517) 355-4676 (see also "Skywatcher's Diary" at www.pa.msu.edu/abrams/diary.html). For just $11 a year you get a two-sided sheet for each month with a superb basic star map on one side and an informative calendar with the sky scene for each night on the other side.

SkyNews (www.skynewsmagazine.com). A Canadian magazine much shorter than *Sky & Telescope* and *Astronomy* and published only bimonthly, but very beautifully illustrated and offering some excellent observational articles.

Online

American Association of Variable Star Observers (AAVSO) (www.aavso.org).

Association of Lunar and Planetary Observers (ALPO) (www.lpl.arizona.edu/alpo).

AstroAlerts. Available from the Sky & Telescope Web site (SkyTonight.com), this is a free service that sends to the subscriber e-mail notices about urgent time-sensitive astronomical events that are in progress. Subjects include solar/auroral activity, comets, variable stars, near-Earth asteroid passages, and more.

Astronomical League (www.mcs.net/~bstevens/al). A confederation of more than 200 amateur astronomy clubs in the United States.

Astronomical Society of the Pacific (www.astrosociety.org). This educational organization, more than a century old, offers a catalog with a wide variety of excellent books, posters, slide sets, software, and CD-ROMs.

Astronomy (www.astronomy.com). The home page of *Astronomy* magazine.

Clear Sky Clock (www.cleardarksky.com). Attila Danko's invaluable predictions of sky conditions (cloud-cover, transparency, "seeing," and much more) for the next two days at thousands of sites in North America.

Comet "chasing" (observation) (www.skyhound.com/sh/comets.html). The Skyhound Comet Chaser's page is full of observing predictions (with updates) of comets currently within the range of amateur telescopes and cameras.

Comets and meteor showers (comets.amsmeteors.org). Gary Kronk's expert information on comets, past, present, and future, and meteor showers.

Eclipse photography and basics (www.mreclipse.com). The ultimate eclipse expert Fred Espenak's beginner-friendly site on eclipses.

Eclipses, authoritative and comprehensive NASA site (sunearth.gsfc.nasa.gov). This site is run by Fred Espenak.

Fireball meteors (www.amsmeteors.org/fireball/fireball_log.html). Reports of fireballs through the year.

Hubble Heritage (heritage.stsci.edu). Offers popular access to Hubble Space Telescope images.

International Dark-Sky Association (IDA) (www.darksky.org). The central bureau for information on light pollution and how to combat it, technologically and legally.

International Meteor Organization (IMO) (www.imo.net).

International Occultation Timing Association (IOTA) (www.lunar-occultations.com/iota).

NASA Spaceflight (spaceflight.nasa.gov). Very detailed ongoing information about the International Space Station and Space Shuttle missions, plus past missions.

Planetary Society (planetary.org). The site of the premier spaceflight advocacy organization, offering vast quantities of infomation about all missions to the planets.

Satellite tracking (www.heavens-above.com). Detailed information on the upcoming days' satellite visibility calculated for the observer's precise location.

Sky & Telescope (www.SkyTonight.com). The home page of *Sky & Telescope* magazine and Sky Publishing Corporation.

Space Weather Bureau (www.spaceweather.com). The latest predictions, information, and images about aurora and solar activity—and about a wonderful assortment of other astronomical objects and events. A great all-around astronomy Web site.

PHOTO CREDITS

Photo credits: pp. 14, 15, 16, 17, Abrams Planetarium © 2005; pp. 18, 23, 37, 48, 74, 75, 102, 118, 196, 213, 220, 231, © Johnny Horne; pp. 35, 62, 113, 140, 141, 144, 157, 165, 181, 182, 183, 184, 186, 187, 193, 210, 211, 218, 223, 225, 238, 255, 259, © Doug Myers; pp. 41, 70, 103, 128, 134, 203, © Ray Maher; pp. 55, 250, © Akira Fujii; pp. 80, 88, 177, 233, 258, © Guy Ottewell; pp. 87, 109, 137, © Steve Albers; p. 112, © Shahriar Davoodian; pp. 116, 147, 245, Chuck Fuller; pp. 121, 163, © Anthony Arrigo; pp. 171, 201, © Nelson J. Biggs Sr.; p. 214, Hubble Space Telescope, courtesy of NASA

Color insert credits: pp. 1, 2, 3, 5, (bottom), 6, 7, 8 (bottom), Johnny Horne; p. 4, Steve Albers, Dennis di Cicco, and Gary Emerson; p. 5 (top), Akira Fujii; p. 8 (top), Richard Yandrick

INDEX

Page numbers in *italics* refer to illustrations.